Easy Veg

Essential know-how and expert advice for gardening success

CONTENTS

PLANNING, PLANTING, AND GROWING 006
WHY GROW YOUR OWN VEG? 008
WHICH VEG TO GROW? 010
WHAT HAPPENS WHEN? 012
WHERE TO GROW 014
RAISED BEDS 016
VEGETABLES IN CONTAINERS 018
PREPARING THE SOIL 020
MAXIMIZING PRODUCTIVITY 022
VEG IN A SMALL SPACE 024
FAMILY GARDENING 026
STARTING FROM SEED 028
STARTING WITH PLANTS 030
STARTING OFF UNDER COVER 032
FEEDING 034
WATERING 036
PROTECTING YOUR CROPS 038
AFTER THE HARVEST 040
YOUR SECOND SEASON 042
EXTENDING THE SEASON 044
COMPOSTING 046
MAKE MORE SPACE FOR VEG 048

SALADS AND HERBS 050
LETTUCES 052
RADICCHIO AND CURLY ENDIVE 054
PEPPERY LEAVES 056
SORREL 058
LEAF CELERY 059
PEA AND BEAN SHOOTS 060
CRESS AND MICROGREENS 061
LEAFY HERBS 062
LEAFY PERENNIAL HERBS 064
TROUBLESHOOTING 066

PEAS AND BEANS 068
PEAS, MANGETOUTS, AND SUGAR SNAPS 070
FRENCH BEANS 072
RUNNER BEANS 074
BROAD BEANS 076
DRYING BEANS 078
TROUBLESHOOTING 080

THE ONION FAMILY 082
ONIONS 084
SHALLOTS 086
SPRING ONIONS 087
GARLIC 088
CHIVES 089
LEEKS 090
TROUBLESHOOTING 092

SUMMER VEGETABLES 094
BUSH TOMATOES 096
CORDON TOMATOES 098
COURGETTES AND SQUASHES 100
RIDGE CUCUMBERS 102
AUBERGINES 104
SWEET AND CHILLI PEPPERS 105
SWEETCORN 106
GLOBE ARTICHOKES 108
TROUBLESHOOTING 110

ROOT VEGETABLES 112
CARROTS 114
BEETROOT 116
RADISHES 117
TURNIPS 118
PARSNIPS 119
POTATOES 120
KOHLRABI 123
TROUBLESHOOTING 124

LEAFY VEG 126
SPINACH 128
CHARD 130
SPROUTING BROCCOLI 132
SPRING CABBAGE 134
KALE 136
ORIENTAL GREENS 138
TROUBLESHOOTING 140

INDEX 142
ACKNOWLEDGMENTS 144

Pack productive plants into your vegetable plot, whatever its size, to give you an exciting range of crops to harvest year-round.

PLANNING, PLANTING, AND GROWING

You can grow your own veg whether you have a large garden or just a few pots on a patio. This section is packed with practical advice to help you plan what to grow when and where, and care for your crops to produce a bumper harvest.

PLANNING YOUR VEG GARDEN

Almost any outdoor space is full of possibilities to a gardener. A balcony or a windowsill is all you need for an edible crop. Take some time to plan how to make the best use of space, whether that is open soil, raised beds, or containers. Choose crops that will grow well in your local climate and that suit your tastes. Find out when you will need to carry out key tasks, notably soil preparation, sowing, and planting, so that you're ready with the seeds and compost you require at the right time. Planning not only leads to bumper crops, but helps to focus your creativity; after all, a veg garden can be every bit as beautiful as one devoted to ornamentals.

SOWING AND GROWING

There are few things more satisfying than picking a crop that you have raised from seed, and for many plants, this is a remarkably easy process. For best results, you'll need to use the right method to sow your seeds, and do so at the right time of year, but even carefree sowings will often perform well. Sometimes, buying healthy young plants can work better than sowing seed, and can give you a head start on the weather. Regular watering and feeding are critical to keep plants healthy and strong, and will help them to shrug off pests and diseases to give you a better crop.

LEARNING FROM EXPERIENCE

Gardeners have always kept notes of sowing and harvest times, successes, and failures, to help them look back at each year and improve their methods. Whatever and wherever you're growing, try doing the same by recording what works and what doesn't, using notes and photos. Reviewing your records will help you modify your activities in the next growing year. Perhaps you were sowing too early, planting a vegetable that doesn't like your soil, or cultivating much more of one crop than you could possibly eat. Reading your notes will also help you develop ideas for expanding your plot or extending the growing season in spring and autumn.

WHY GROW YOUR OWN VEG?

Growing veg has surged in popularity in recent years. People have become more aware of what they eat and where it comes from, and have rediscovered that fresh produce is far superior to anything available in the shops. Tending your own plot is easier than you might imagine, especially if you banish ideas of total self-sufficiency and start with the simple aim of enjoying seasonal vegetables picked in their prime.

Vegetables are richest in nutrients when eaten freshly picked. There's no substitute for home-grown crops.

EAT FRESH

Vegetables taste better – and are better for you – when they are plucked straight from the garden. Leaves are juicy and crisp, roots have distinctive earthy aromas, and natural sugars bring a fleeting sweetness to such vegetables as peas and sweetcorn that is absent from shop-bought produce. Home-grown vegetables also have superior flavour because they can be picked when very young or at the perfect moment of ripeness – feats impossible with commercial crops that are harvested in bulk, stored, transported, packaged, and displayed on a shelf before they reach your table. Growing your own gives you a supply of irresistible vegetables, perfect for outdoor snacking or cooking up seasonal dishes in your kitchen.

TOP TIP GET TOGETHER WITH A GROUP OF FRIENDS TO SWAP SEEDS, PLANTS, PRODUCE, TOOLS, AND IDEAS. IT'LL HELP SAVE MONEY AND MAKE GROWING YOUR OWN EVEN MORE FUN.

Lush salad leaves burst with flavour and crunch when freshly picked, and can be used to create colourful displays.

TAKE CONTROL

Growing your own veg gives you total control over what you eat. You know exactly how your crop has been cultivated and what you have applied to it. Most gardeners choose not to use pesticides on edible plants and instead opt to grow organically, which involves using barriers and companion plants to keep pests away from crops, trying to encourage beneficial wildlife into the garden, and accepting a few nibbled leaves here and there. Planting vegetables that suit the local climate and choosing disease-resistant varieties also helps to keep plants healthy.

Growing your own lets you choose varieties with the taste, colour, and texture that you like, and gain access to crops that are difficult or expensive to buy, such as herbs, heritage carrots, or fresh kohlrabi.

The range of varieties available from seed far exceeds the choice in a shop.

Make gardening a family activity. It motivates children to get into the fresh air and can spark a lifelong passion for plants.

Composting garden waste reduces the amount going into landfill. Compost bins can be bought or easily made at home.

WASTE LESS, LIVE BETTER

Growing your own food helps you to re-establish a meaningful connection with nature, the soil, and the seasons. This is not only an effective antidote to the frantic the pace of modern life but of huge benefit to the environment. The simplicity of picking what you need when you need it means that you are always working with rather than against nature. Packaging is virtually eliminated, and "food miles" – the distance the food you eat is transported from leaving the farm to arriving on your plate – fall away. And because you only harvest crops when you need them, you automatically cut down on food waste; what's more, making use of seasonal crops often inspires greater creativity in the kitchen. Any organic waste that you do produce – vegetable peelings and spent crops – can be easily composted and recycled into rich, crumbly compost to spread on the soil the following year (*see pp.46–47*).

HAVE FUN

Whether you're aiming to fill a few pots with herbs or produce a large-scale veg patch, growing your own is an enjoyable experience. Browsing seed catalogues opens up a whole world of varieties to try – some weird and wonderful, and some that you may simply never have thought to grow. Allow yourself time to daydream about the possibilities and have fun drawing up a plan. Get kids involved in choosing what to grow too, so they have their own plants to care for, harvest, and eat. The pleasure of sowing seeds, nurturing seedlings, and watching plants flourish is obvious. Of course, sometimes things go wrong too, but failures are quickly forgotten when you can unearth, pick, or cut your own fresh produce.

WHICH VEG TO GROW?

The best vegetables to grow are those that you like to eat and are most important in your kitchen. Many factors will influence your choice: some crops – such as newly dug potatoes or young broad beans – simply taste incomparably better than their shop-bought equivalents; others, such as salad leaves and aromatic herbs, give you a renewable supply of fresh flavours for the table at a fraction of the cost of buying them in a supermarket. You'll soon discover what grows successfully in your garden, but when starting out, it's best to begin with tried, tested, or locally recommended varieties.

Hanging plant pockets over a balcony provides useful vertical growing space.

HOW MUCH SPACE IS THERE?

There is a lot that you can do make the most of a small plot (*see pp.24–25*), but be realistic about what you can fit in. Some crops take up a lot more room than others: for example, many salads are compact and high-yielding, while globe artichokes form large plants that crop only for a short time. This doesn't mean that larger plants can't be grown in smaller plots, or even pots, but it pays to prioritize the most productive crops. Measure your plots or raised beds and sketch a plan to ensure that you give crops enough room to grow to maturity. Containers too can provide new ways to squeeze in more edible plants.

WHAT WILL THRIVE IN YOUR GARDEN?

Growing the right plant in the right place is a guiding principle in gardening. Local climate has a huge bearing on what will flourish; heat-loving tomatoes and aubergines won't do well outdoors in cool regions, while tender salad leaves will struggle in areas with hot, dry summers. Choose varieties bred to suit local growing conditions, and ask your neighbours, friends, and local allotment gardeners what grows well. The aspect of your plot will affect the amount of light available to plants – most veg plants do best in sunny positions. Consider the soil in your garden: some vegetables prefer sandy soils, while others thrive in heavier soils containing more clay.

TOP TIP PLANT ROOT VEGETABLES IN WINDY SITES, WHERE ABOVE-GROUND CROPS COULD BE DAMAGED.

Clay soils hold moisture and are slow to warm, but crops such as cauliflower 'Romanesco' and Calabrese thrive in such heavy soils.

HOW MUCH TIME DO YOU HAVE?

Growing veg should be a relaxing pursuit. If you have lots of other commitments, start small and gradually expand what you grow as you discover what works for you. Tending to plants little and often is the best approach as it keeps tasks such as weeding manageable, and prevents gardening from becoming a chore. Remember that plants grown in containers and under cover need regular watering, especially in summer; those grown in the ground are much more able to look after themselves. Also, consider your holiday plans; there's little point growing lots of summer crops only to find that you miss them while you're away.

Leaf beet (chard) is an ideal plant for a busy gardener because its tender leaves regrow after cutting.

PLANNING FOR A YEAR-ROUND HARVEST?

Your garden can keep you fed and active all year round if you look beyond the obvious late spring and summer crops. Winter and early spring vegetables, such as parsnips and sprouting broccoli, will take up significant space in your garden for many months, but are a wonderful harvest when there is little else growing outside. You can extend your harvesting period over the summer by making small sowings of salad leaves and root crops every few weeks, and raising young plants in small pots, ready to instantly fill any gaps when another crop is harvested.

NEED TO KNOW

If you are new to growing vegetables, start with these easy-to-grow crops:
Salads • Radishes • Potatoes • Herbs • Peas • Broad beans • Garlic • Swiss chard • Kale • Tomatoes

Kale 'Nero di Toscana' (cavolo nero) can be picked as required throughout winter and is hardy enough to withstand hard frosts.

Bring on plants in pots to replace crops you harvest from the ground.

WHAT HAPPENS WHEN?

Successful vegetable gardening is all about good timing. The guide below sets out the jobs that you should be tackling throughout the year, but be aware that timings will vary quite considerably according to the local climate and the changing weather conditions from year to year. Recognizing the perfect moment to sow, plant, harden off, and harvest gets easier with experience. The best way to learn is to start growing by sowing small quantities of seed every two weeks to see how they perform; an early start doesn't always produce the quickest crop.

SPRING

This is the season for sowing, but don't start too early when the soil is still cold. Begin with hardy plants, such as broad beans and peas, then sow lettuces, radishes, and spring onions (these may need some early protection under cloches). Sow heat-loving chillies and tomatoes and on a warm windowsill.

From mid-spring, make successional sowings of crops such as carrots and spinach, and plant out seed potatoes and onion sets. Weed regularly, thin out seedlings, and add supports for climbing peas. In late spring, earth up potatoes, harden-off seedlings raised under cover before planting them out, and sow summer crops, such as runner beans. Early sowings should be ready to harvest as summer approaches.

KEY TASKS

- Begin sowing hardy crops, such as beetroot and peas, outdoors.
- Plant potatoes, and onion and shallot sets.
- Weed regularly.

Radishes may be one of the first crops ready for harvest in late spring. Pull aside the leaves and harvest roots that are over 2cm (¾in) in diameter.

Liquid fertilizer ensures plants receive all the nutrients needed for healthy growth.

SUMMER

By early summer you will begin to reap the rewards for your work in spring as peas, broad beans, new potatoes, baby beetroot, and many other vegetables become ready to harvest. Pick them regularly and often to enjoy crops at their best and to free up space for more successional sowings of salad and root crops; these will let you keep harvests going into autumn.

Tender crops, such as squashes, tomatoes, and sweetcorn, can be sown outdoors in early summer or planted out once they have been hardened off. Now's the time to think further ahead too: planting out leeks, kale, and sprouting broccoli in early summer will give you valuable crops during winter and early the following spring.

Keep vegetables well watered in hot summer weather; plants growing in containers may need watering daily and will also benefit from regular feeding with a liquid fertilizer.

KEY TASKS
- Sow and plant tender crops outdoors.
- Harvest a wide range of veg.
- Water regularly, especially if plants are growing in containers.
- Keep sowing successionally for continuous crops.

AUTUMN

Early autumn is a bountiful time, as tomatoes, squashes, onions, and root crops are waiting to be harvested. Find ways to store any gluts for use through the colder months. In warmer regions there is still time for a last sowing of many vegetables in early autumn. Try autumn sowings of broad beans and hardy peas to overwinter, as well as planting garlic and spring cabbages. When each crop comes to an end, uproot the spent plants and add them to your compost heap.

As nights turn colder, protect late sowings of carrots, salads, and French beans from frost with cloches and add a cosy layer of mulch around globe artichoke crowns and overwintering crops. Mulch bare soil with a thick layer of compost before winter.

Harvest squash before the first frosts, cutting the stalk away from the fruit.

KEY TASKS
- Plant garlic and spring cabbages.
- Harvest squashes, onions, tomatoes, and root crops.
- Tidy up and compost waste.
- Mulch around overwintering crops.

WINTER

The coldest months are a quiet time in the garden, but there is still plenty to do. Hardy crops, such as leeks, parsnips, and kale, can tough out the coldest weather and will be there to harvest for hearty soups and stews throughout winter. Spend the darkest days studying seed catalogues, planning what to plant in the next growing season and ordering seeds, seed potatoes, and onion sets. Start chitting seed potatoes (see *p.120*) on a cool, bright windowsill in late winter, when garlic and broad beans can also be planted in well-drained soil outdoors or in pots. As spring approaches, dig manure into your beds if the soil isn't too wet.

KEY TASKS
- Plan what to grow in the coming year.
- Harvest your winter veg.
- Plant garlic and sow broad beans.
- Chit seed potatoes.

Winter cabbages are hardy and can remain in the ground even through frosts until you choose to cut them.

WHERE TO GROW

Most common vegetables are annual plants. To reach maturity and produce a good crop, they need to do a lot of growing over a short season and so demand plenty of sunlight and water, as well as good soil to sink their roots into. Given these conditions they will grow almost anywhere – in open soil, raised beds, or containers – giving you huge flexibility in how you lay out your garden. You should avoid siting plants where they are likely to be overshadowed by trees or shrubs, and be aware that some taller plants, such as cucumbers and tomatoes, are susceptible to wind damage, and so prefer a sheltered spot.

IN THE GROUND

If the soil in your garden is crumbly and free-draining, consider yourself very lucky: make full use of this valuable resource to plant edible crops. You can plant veg in neat rows spread over broad beds, but this requires a large space and leaves lots of bare soil to weed. It may be more practical to create narrow beds, divided by paths, as this removes the need to walk on the cultivated soil and means that crops can be planted more densely, crowding out any weeds. Beds can be square, rectangular, curved, or circular to suit the shape and design of your plot.

COLD AND WIND Avoid planting vegetables in areas where cold air tends to pool because they will be more susceptible to frost damage (*see right*). Think about erecting hurdles or even growing low hedges on the windy side of your veg plots; they will serve as wind diffusers, preventing damage to delicate plants.

MIXED PLANTING If you don't want to create a dedicated vegetable plot, try adding annual flowers and herbs to create a potager-style garden. Edible plants can also be used to add bold foliage to existing flower borders by sowing in situ or using young plants to fill any gaps.

Growing in open soil particularly suits perennials that regrow year after year and plants like kale that prefer deep, fertile soil.

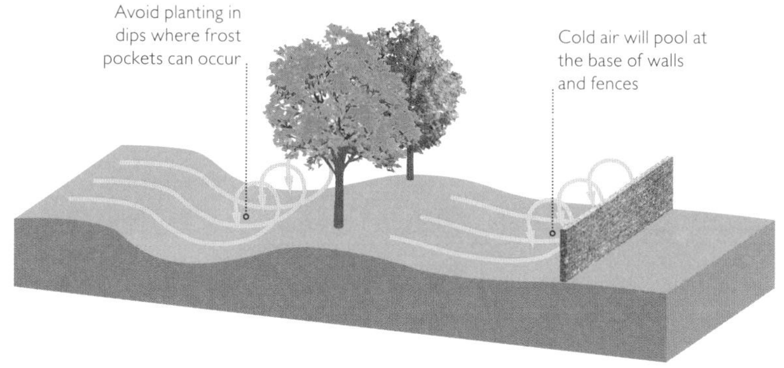

Cold air is denser than warm air and will tend to "pool" in hollows and at the bases of walls and fences.

RAISED BEDS

If you have poor garden soil (or none at all), raised beds are an excellent option. Straightforward to construct from kits or a variety of other materials (*see pp.16–17*), they can be built to any length to suit your plot, and filled with a good-quality topsoil and compost mix. They can be positioned on soil, paving, or concrete and are easy to maintain because they are raised above ground level. This makes them easy to plant up, and quick to weed and mulch with compost; furthermore, the soil they hold tends to drain well after heavy rain and warms up faster than the open ground in spring, making it excellent for early sowings. Raised beds can be packed with plants to maximise yield and arranged in groups to make an attractive feature with paved or grass paths in between.

Dense plantings flourish in raised beds because all maintenance can be done from the path without standing on the soil.

CONTAINERS

Growing veg in containers and hanging baskets is a great solution if you only have a patio, paved courtyard, or balcony. Pots have a place in a larger garden too, because they can be moved as needed (shifting lettuce into shade for summer, for example) - and rearranged to form interesting displays through the seasons. Opt for the biggest containers available, because they will hold plenty of compost and won't dry out too rapidly, which is important because pot-grown plants rely on you for regular watering and feeding. Be imaginative and upcycle old household containers to add a touch of quirkiness to your planting.

A range of crops, including cavolo nero, grow well in containers.

UNDER COVER

Access to growing space that is warm, light, and sheltered during spring is extremely valuable for giving your first sowings of seeds a head start while the soil outdoors is still cold. If you live in a cool climate, a sheltered space allows you to raise tender crops, like tomatoes, without risking frost damage and provides them with the heat they need to ripen during summer. If you are lucky enough to have a greenhouse then make full use of it, but a sunny windowsill works just as well as a nursery for vegetable seedlings, or even for fully-grown heat-loving crops, if you can spare the space.

Make early spring sowings on a windowsill to plant outdoors as the weather warms.

NEED TO KNOW

- Most vegetables flourish in full sun; avoid planting in the shade of hedges, trees, and buildings.
- Free-draining soil is essential; never sow in waterlogged soil, where water puddles after rain.
- High winds will damage delicate leaves; provide shelter in gardens exposed to strong winds and beware of wind whipping between buildings.

RAISED BEDS

You may choose to elevate your vegetables from ground level and grow them in raised beds. Doing so has many advantages: access is easy, making tasks such as planting, weeding, and watering less taxing. Raised beds work well with "no dig" cultivation (*see pp.20–21*) saving time and energy; and though construction requires more effort at first, raised beds are much less work to maintain than open soil.

Defined edges and tidy paths make raised beds a neat option for compact city gardens..

SIZE AND SHAPE

Raised beds allow you to grow your own veg almost anywhere. They are ideal for city gardens, paved courtyards, sunny patios, and especially the gardens of new-build houses where the soil is often badly drained or full of rubble.

Garden centres sell self-assembly kits that can be put together in minutes; alternatively, you can build your own raised beds from inexpensive materials with some basic tools and DIY skills.

The beds can be almost any length, but should ideally be no wider than around 1.2m (4ft) so that you can reach across them comfortably without standing on the soil. Aim for a depth of at least 30cm (1ft), which is sufficient to support most veg crops; the deeper the soil the more it will hold on to moisture, reducing the watering frequency required.

The higher the sides of the bed, the less you'll need to bend over when tending to your crops – a big plus if you have limited mobility.

Raised beds drain more freely than open soil, and warm up more quickly, so can be planted earlier in the season.

SITING AND PREPARING

Choose an open, sunny spot for your raised bed. Make sure the bed can be easily reached with a hose or is close to a water supply as you'll need to water regularly in the summer months. The larger the bed, the more soil you'll need to fill it; bought soil is usually delivered in bulk in large bags, so think carefully about how you will transport it to the site.

Before positioning the bed and filling it with soil, carry out a little ground preparation. Remove weeds from the site, level the ground, then loosen the soil below with a garden fork to aerate it and improve drainage. If you are placing the bed on concrete or compacted soil, it will need to be at least 30cm (1ft) deep to let the roots of your veg plants grow freely.

Regular summer watering is essential, so site your bed near a water supply.

BUILDING A BED

Make a sturdy raised bed out of a material that will last. This frame is built from plastic decking boards, but the same techniques work for a wooden frame. A depth of 30cm (1ft) is created by making the bed two boards deep.

YOU WILL NEED A saw, screws, a mallet, an electric drill and screwdriver and – ideally – someone to help.

1 Cut four 1.2m (4ft) planks for the width of the bed and four to your chosen length measurement. Cut four 60cm- (2ft-) long corner posts, plus (for beds over 2m (6ft) long) two posts to position every 1m (3ft) along the length of the bed for added stability. Lay the boards out in their intended positions.
2 Dig out any large stones or protrusions in the ground where the frame will rest in order to make the bed lie as level as possible.
3 Stand the bottom row of boards on their edges to form a rectangle and use a set square to ensure each corner is a right angle. Drill pilot holes through one side of the long board, then screw the long and short boards together to create a rectangular frame.
4 Hammer the four corner posts into the ground. The top of each post should be just below the top of the bed when the top row of boards is added.
5 Screw the lower boards onto the corner posts to secure the framework. Next, add the top row of boards, screwing this to the posts too.
6 Hammer in posts to support the long sides of the bed as required. Screw the boards to these lateral supports too.
7 Fill the bed with a mixture of good topsoil and peat-free compost. Allow the soil to settle for two weeks before planting.

VEGETABLES IN CONTAINERS

Containers provide space for growing vegetables wherever it is needed, and let you scale up your production easily from one year to the next. A huge range of crops, including root veg such as carrots and potatoes, can be grown in containers, turning a balcony or a patio into a living larder. The limited amount of compost that a container can hold means that plants need more feeding and watering than those in the soil. However, there are advantages too: pots rarely require weeding, and growing in containers provides a degree of protection from common pests and diseases.

CONTAINER ESSENTIALS

Most vessels that can hold compost can be used for growing vegetables. Make sure, however, that the container has drainage holes in its base to allow excess water to escape, because plants will rot and die in waterlogged soil. Match the size and depth of the container to your chosen crop: large, hungry plants, such as potatoes, courgettes, and climbing beans, will need a pot at least 50cm (20in) wide and deep, but shallow-rooting salads, spring onions, and radishes can thrive in compost just 10cm (4in) deep. Smaller pots will need more frequent watering than larger containers. The material the container is made from also affects how often it will need watering. Terracotta pots are porous and lose moisture through their sides, and metal containers will heat up quickly in direct sunlight, so both will need watering more often than plants in plastic containers.

Certain materials including lead, asbestos, some plastics, and treated woods may be toxic to plants or people. In short, if you are not certain that a container is safe, don't use it.

Containers can quickly bring plain or neglected parts of your garden into productive use.

CREATIVE DISPLAYS

By matching materials, finishes, and colours, you can quickly give your container garden a distinctive look. Group pots together to increase their impact and try including a few taller plants, such as sweetcorn and runner beans to add vertical interest. Vibrant flowering plants can be combined with the flowers and fruits of vegetables to enhance the display and help attract pollinating insects. Add an extra dimension by using hanging baskets, or attaching pots to walls or fences.

Creative stacking of containers can increase your effective growing space.

POSITIONING POTS

Most vegetable plants need plenty of warmth and light to thrive; as a general rule, position containers in the sunniest spots available. Leafy vegetables, however, prefer a little shade during hot weather, as do plants growing in smaller pots that dry out rapidly. Leave sufficient space between pots for plants to flourish and make sure that taller plants don't cast shade on shorter crops. If your containers aren't too heavy to move, tweak your display periodically to keep it looking fresh and move tender plants under cover to keep them cropping as the weather turns colder. Tall or bushy plants can be blown over easily by the wind; move them to a sheltered position. Plant them in large, heavy containers filled with soil-based compost and keep them well watered.

TOP TIP MOVE TENDER PLANTS LIKE CHILLIES AND AUBERGINES INDOORS AS THE WEATHER COOLS IN AUTUMN.

Heat-loving crops, such as peppers and tomatoes, ripen best given a south-facing position throughout summer.

CROP PROTECTION

Container-grown veg experience fewer problems with pests and diseases than those grown in the ground. Fresh compost is sterilized to make it free of the soil-borne pests and diseases that can be the scourge of established veg plots. Crops in pots tend to be grown in small scattered groups, and need little thinning, which make them less likely to attract insect pests than densely sown garden rows. Containers on balconies and otherwise raised above ground level are also less vulnerable to attacks by slugs, mice, and carrot flies. Plants can be protected from damage by pigeons and insect pests by covering the container in fine insect mesh or fleece.

Copper tape around the circumference of a pot is a barrier to slugs and snails.

CHOOSING COMPOST

Don't use garden soil to fill containers because it will certainly harbour weed seeds and soil-borne pests and diseases. Instead, choose a peat-free, multi-purpose compost. This is made from composted bark, coir, wood chips, or green waste, and is relatively cheap and available from any garden centre. Avoid composts that contain large lumps of woody or fibrous material, as these make it difficult to sow seeds evenly. Keep multi-purpose compost consistently moist with regular watering because it's difficult to rewet once it dries out.

For longer-term plantings of perennial vegetables and herbs, use a heavier soil-based compost. This is made from soil, coarse sand, composted organic material and fertilizer, which doesn't break down over time like multi-purpose compost.

To use home-made garden compost in containers, mix it 1:1 with topsoil or leafmould. It is however, unsuitable for sowing seeds as it is too high in nutrients and may carry fungal diseases.

Choose a compost that suits your crops and types of container.

PREPARING THE SOIL

The soil in your garden is a valuable resource that needs to be looked after and improved year on year. Vegetables thrive in a wide range of soils, but perform best in those containing a high proportion of organic material (compost). This supplies the nutrients essential to support rapid growth and heavy cropping, and to help plants bounce back from attacks by pests and diseases. Healthy soils also drain well after heavy rain but retain moisture during dry weather, which reduces the need for watering. Get to know your soil and make the most of its potential.

IMPROVING YOUR SOIL

Get to know your soil by picking up a handful and squeezing it: a clay soil will clump together and feel sticky and smooth while sandy soil will feel dry and run through your fingers. Clay soils are often poorly drained and may bake hard in summer, but retain moisture and nutrients well; sandy soils are often dry and can drain quickly so nutrients wash away, but they are easier to work.

The best way to improve either type of soil is to add large amounts of organic matter – compost, recycled green waste, or rotted manure – every autumn. Organic matter can be dug in to the soil, or simply be left on the surface and planted into directly. Whether you start with a clay or sandy soil, the addition of organic matter will result in a rich, dark brown growing medium with a crumbly, open structure that is full of earthworms and other beneficial soil organisms.

Combine soil (left) and well-rotted compost (right) to provide vegetable plants with an ideal nutrient-rich, free-draining growing medium.

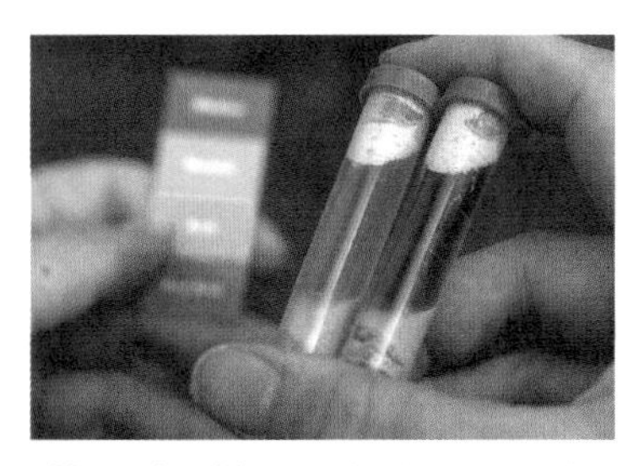

pH testing kits are simple to use and help you understand your soil.

SOIL ACIDITY The ideal garden soil is slightly alkaline, with a pH of about 7.5 (where 7 is neutral). However, many soils are acidic; these conditions limit the availability of key nutrients to plant roots and affect the activity of beneficial soil bacteria, lowering soil fertility.

You can use a simple soil testing kit (available from your garden centre or online) to measure the pH of your soil; no special skills are required. If you discover that your soil is acidic, you can apply pulverized limestone as a neutralizing remedy. Carefully follow the instructions given with the product because adding too much lime will have its own negative effects.

Lime is usually spread in winter so it has time to take effect before planting.

DEALING WITH CLAY SOIL

Clay soils (or "heavy" soils) can be more difficult to deal with than sandy ones because their structure is easily damaged if they are worked or walked on when wet. Adopting a "no dig" regime (*see below*) is the easiest way to improve clay soils but if you do choose to dig them, do so in the relative dry of autumn, lifting big clods that will be later broken down by winter frost. Adding sand or grit to lighten clay soils is rarely worthwhile; focus on opening up soil structure and improving drainage with the regular addition of organic matter.

Stand on wooden boards to avoid compressing clay soils as you work.

TO DIG OR NOT TO DIG?

For generations, gardeners have advocated digging soil every year to aerate it, open up its structure, and incorporate nutrients. However, over-digging does not deliver these benefits but actually disrupts the natural balance of particles, space, and beneficial microbes in the soil and produces an unstable footing for plant roots. Digging also brings buried weed seeds to the surface to germinate. Many gardeners today prefer "no-dig" methods that do away with this time-consuming and physically exhausting (especially with heavy clay soils) task.

Whichever method you use, you should apply a 5cm- (2in-) thick layer of organic matter (well-rotted compost or manure) to the soil every autumn. If you adhere to traditional garden practice, dig it in thoroughly to a spade's depth; however, if you follow the no-dig system, simply leave it on the surface to be incorporated into the soil by earthworms and broken down by microorganisms. The mulch itself inhibits weed growth and acts as an insulating layer, keeping the soil warmer.

The no-dig method is ideal for enriching the soil.

NEED TO KNOW

- If you're not using home-made compost on your beds, source your organic matter carefully to ensure it is not contaminated with herbicides.
- Apply lime (if needed to reduce acidity) in the winter, so that it has time to take effect before the next year's planting.
- Always remove weeds before they set seed to prevent them spreading around your plot.

WEEDING

Weeds compete with your crop plants for water, light, and nutrients, and can harbour pests and diseases. When establishing a new veg bed it is essential to dig out all traces of perennial weeds, before adding a thick layer of organic matter to the soil (and either digging it in, or not). Thereafter, remove weeds regularly with a fork or hoe; annual weeds, such as hairy bittercress and chickweed, will spring up from seed, while perennial weeds, like dandelions and couch grass, spread via their fleshy roots as well as seeds. Weeding is hard work, but becomes easier with time as you reduce the sources of weed seeds.

Loosen soil with a fork to ease the removal of the roots of perennial weeds.

MAXIMIZING PRODUCTIVITY

Some gardeners take pride in perfection, aiming for ordered rows of pristine veg; others are more relaxed, accepting a little chaos in exchange for convenience. But almost all gardeners want to boost the productivity of their plots to maximize the return on the investment they have made in preparing their beds and containers and improving their soil. This book – and your growing experience – will provide a guide as to what each plant needs to thrive and deliver a good crop, but there are also some clever techniques that will increase your garden's yield throughout the year.

KEEP THE SOIL COVERED

Beginner gardeners will typically sow in spring, harvest in summer, and then find their beds are largely empty for the rest of the year. To avoid this pitfall, do a little planning: work out when each of your spring sowings will be ready to harvest and what vegetables could follow on from them. Be ready to sow the next crop as soon as the first is harvested; alternatively, buy or raise young plants in pots or under cover to instantly fill gaps, so reducing the time the second crop takes to mature. This will give you plenty to harvest during winter and early spring. The soil can be improved later in the year by applying a mulch around crops in autumn, leaving space for air to circulate at the base of each plant.

Filling a bed so that no bare soil is visible maximizes productivity while keeping weed growth in check.

SUCCESSIONAL SOWING

This simple technique involves sowing small amounts of seed repeatedly – usually every 2–3 weeks – to produce a continuous harvest over a long period. It is often overlooked by beginners, but suddenly makes sense once you have experienced trying to eat your way through a glut of a single vegetable.

Successional sowing also has the advantage that if one sowing fails (perhaps because it was made too early or the soil was too dry) then all is not lost. Successional sowing is also a good way to learn and see for yourself how different conditions affect the speed of germination and growth.

Make your next sowing when the previous row of seedlings germinates

TWO-IN-ONE CROPS

You can raise two crops at the same time in one area by intercropping – combining veg that take different times to mature or have complementary growth habits (*see right*). For example, a fast-growing crop can be used to fill bare soil between plants that are slow to germinate or grow, and it will be harvested well before it begins to interfere with the growth of the second crop. Try sowing radish seeds thinly between clusters of parsnips, or raise cut and come again salads among large plants, such as sprouting broccoli.

In a similar way, underplanting makes use of the soil beneath tall crops to grow low or sprawling vegetable plants. Fill the gaps between young climbing beans and peas with salads or turnips before they cast too much shade, or grow sprawling squashes at the base of slender sweetcorn plants. Both of these intense growing methods require fertile soil and may need extra watering, but they reduce the need for weeding.

INTERCROPPING COMBINATIONS

SLOW	FAST
Garlic	Radishes
Leeks	Kohlrabi
Kale	Baby beetroot
Onions	Loose leaf lettuce

UNDERPLANTING COMBINATIONS

TALL	SHORT
Cordon tomatoes	Basil
Sprouting broccoli	Radicchio
Sweetcorn	Dwarf French beans
Climbing beans	Spinach

Quick-growing loose-leaf lettuce has been planted here between slower-growing parsnip plants.

CROP ROTATION

Crop rotation is a traditional practice in which related vegetables are not grown in the same part of the garden each year but instead moved ("rotated") from one year to the next. This helps to prevent the build up of soil-borne pests and diseases, and helps rebalance the availability of nutrients in the soil.

Rotations often follow a three- or four-year cycle, in which plants from the cabbage family (brassicas), peas and beans (legumes), root crops, and potatoes and tomatoes (which come from the same plant family) are moved to a new location each year. Other crops, such as salads, can be fitted in around them. Rotation is easy to plan if you have several beds, but even in small gardens it makes sense not to grow the same crop in the same place year after year.

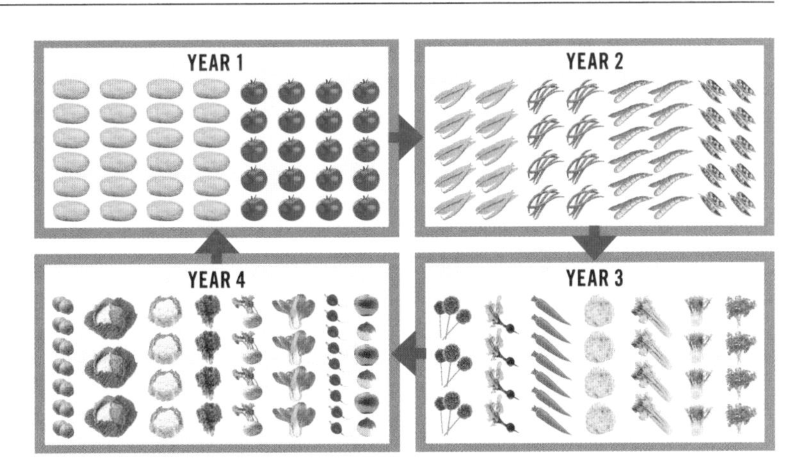

The rotation you use will depend on the crops that you wish to grow. Here is a sample four-year rotation.

YEAR 1 Potatoes • Tomatoes
YEAR 2 Legumes • Broad beans • French beans • Peas • Runner beans
YEAR 3 Roots • Alliums • Beetroot • Carrot • Celeriac • Celery • Florence fennel • Parsley • All other root crops
YEAR 4 Brassicas • Brussels sprouts • Cabbage • Cauliflower • Kale • Kohlrabi • Oriental greens • Radish • Swede • Turnip

VEG IN A SMALL SPACE

A surprising number of plants can be packed into a small space to create a haven of green growth that will provide generously for your kitchen. Make efficient use of any open soil you have, be creative with containers, and exploit the vertical space above your plot. A veg garden needn't be dull if you combine plants for visual effect, just as you would ornamentals. Leaf crops, such as Swiss chard and kale, provide bold colour and texture; peas, ridge cucumbers, courgettes, and herbs give spring and summer flowers; and fruiting crops, like tomatoes, chillies, and squashes, add a burst of colour as they ripen.

OPTIMIZE OPEN SOIL

If your garden contains open soil or raised beds, start by improving your valuable soil resource by adding well-rotted compost (*see pp.46–47*) to increase fertility and moisture retention. This will significantly boost the yield of your plot. Plant your vegetables in blocks of short, staggered rows rather than single long rows – you'll fit more plants into a bed. However, don't be tempted to plant more densely than recommended – yield will fall and your plants will be more vulnerable to disease. Plan your plots carefully before planting, positioning taller plants where they won't cast shade on smaller crops.

Raised beds are an efficient way to grow veg in a small space.

USE VERTICAL SPACE

Make full use of your garden's three dimensions by thinking vertically. The easiest way to grow upwards is to plant climbers, such as French beans or squashes, up supporting poles, trellises, or arches. Tall crops not only add structural interest to your garden, but they can boost yield because they are able to access valuable sunlight not available at lower levels. They are also useful as screening (hiding bins or compost heaps) and can clothe a blank wall or dull fence. Another way to exploit your garden's vertical space is by planting in containers elevated on any stable support, such as a windowsill, a plinth, or – fashionably – a ladder leant against a wall. Containers can be made from a variety of cheap or repurposed materials, such as old pots and pans, half-tyres, or wooden crates – just ensure that they have drainage holes and are made from non-toxic materials, free from wood preservatives.

Even a small patio area can be productive when surrounded by climbing and trailing veg plants.

CONSIDER ASPECT

City gardens are likely to be surrounded on one or more sides by buildings or trees that will affect local patterns of light, rainfall, and wind. Watch to see where light falls in your garden over the course of a day, and – as a general rule – give vegetable plants the sunniest spots available. Try painting walls or fences white to reflect light and brighten up the shadier areas. Choose heat-loving crops, such as tomatoes, peppers, sweetcorn, climbing courgettes, and basil for sun-baked, south-facing spots and plant shade-tolerant leafy vegetables, such as Swiss chard, lettuce, and spinach in the shadier areas. Walls can prevent sufficient rain from reaching plants on their lee side, so ensure these areas are adequately watered.

A lick of white paint on a wall helps to reflect light back into the vegetable garden.

TOP TIP GAPS BETWEEN WALLS, FENCES, AND BUILDINGS CAN CHANNEL AND INTENSIFY WINDS THAT DAMAGE TALL PLANTS. TRELLIS OR MESH PLANTED WITH CLIMBERS WILL HELP TO DIFFUSE POWERFUL GUSTS.

NEED TO KNOW

The best crops for small spaces:

- Quick-cropping: radishes, salad leaves, oriental greens, spring onions, herbs.
- Dwarf or bush habit: chillies, aubergines, French beans.
- Highly productive: courgettes, potatoes, tomatoes.
- Climbing: runner beans, peas, cucumbers, squashes.

MAKE A PALLET GARDEN

With some basic DIY skills, a length of landscape fabric, and a staple gun (or some roofing nails and a hammer) you can transform an old wooden pallet into a versatile vertical planter. Pallets vary in shape and size: choose one with large gaps between the slats or use a hammer to knock out alternate slats. Lie the pallet flat on its front, then staple or tack landscape fabric along the backs of the horizontal struts of the pallet. Allow sufficient slack in the fabric to make a deep loop at each slat. This loop will form a pocket that holds soil at the level of each slat. Next, staple the fabric to the vertical sides of the pallet to close the pockets at the sides. Position the pallet against a south-facing wall, and secure it firmly to the wall with screws or cables. Ensure that the planter is level and stable before filling the pockets with a multi-purpose compost.

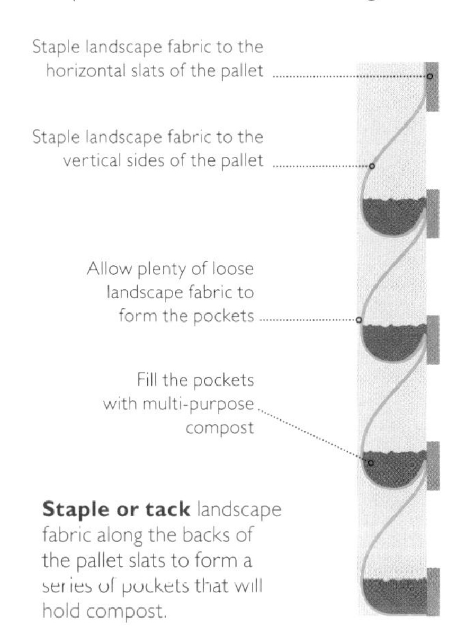

Staple or tack landscape fabric along the backs of the pallet slats to form a series of pockets that will hold compost.

Shallow-rooted vegetables, such as lettuces, radishes, and bok choi are ideal choices for a pallet planter.

FAMILY GARDENING

Most kids need little encouragement to help out in the garden. Digging, planting seeds, watching them grow, and harvesting tasty crops are fun activities that are also enriching and educational. And children, with their naturally imaginative, can-do approach, usually make successful gardeners. Given their own small plot, they will often develop a sense of pride in their outdoor work that can grow into a lifelong love of gardening. Choosing crops that are easy and quick to grow, and that produce a generous harvest, is key to maintaining kids' interest over the long growing season.

GETTING KIDS INVOLVED

Before the growing season, encourage your children to look at gardening books or websites to choose what they'd like to grow. They're far more likely to remain engaged if they have invested in the process from the start. Find them a space – even if it's just a good-sized container – for which they can take full responsibility and let them decorate their patch as they like. Younger children love having their own tools, so it's worth investing in a small trowel, hand fork, and watering can which are easy for them to use – you also won't have to worry about them losing yours.

Small watering cans are light enough for kids to fill, carry, and use.

CUSTOMIZE CONTAINERS

Container gardens are an easy win. They are quick to plant, simple to maintain, and can be moved into shelter in case of bad weather. Using quirky or unusual containers just adds to the fun: try colanders, laundry baskets, crates, or even old wellington boots. Line gappy containers with the thick plastic from an old compost bag and always make drainage holes in the bases of pots where there aren't any already. Children understand the importance of reusing and recycling, so have a go at sowing seeds into toilet roll tubes, egg boxes, and yogurt pots.

Egg boxes and toilet rolls make ideal, biodegradable seed trays.

Starting from seed lets children witness the whole plant life cycle. Grow the seedlings indoors and move into pots outdoors when ready.

FUN CROPS TO TRY

Introduce your kids to gardening with quick-growing vegetables, such as radishes, turnips, rocket, spinach, and pea shoots, so that there isn't too long to wait between sowing and harvest. Climbing crops, like cucumbers, French beans, and runner beans, race up their supports fast once they are established and are fun to pick; and naturally sweet vegetables, such as cherry tomatoes, sweetcorn, peas, carrots, and beetroot have obvious appeal to children. Courgettes are usually a hit too because it's easy to get a huge crop for very little effort – one plant per child is enough. The colourful fruits of winter squashes swell rapidly through summer and smaller varieties hang like baubles when the plants are grown as climbers. Potatoes are a must-grow crop too, because they're really easy in open soil or a large container and unearthing baby new potatoes is just like digging up buried treasure.

TOP TIP CHOOSE CROPS WITH LARGE SEEDS THAT ARE EASY FOR KIDS TO HANDLE, AND ONES THAT GERMINATE QUICKLY AND RELIABLY TO PROVIDE FAST RESULTS.

Radishes sprout within a few days of sowing, are ready to harvest in a month or so, and grow through spring and summer.

Themed plantings give kids a clear sense of purpose in their gardening.

CREATE A PIZZA GARDEN

Establish an instant connection between plants and delicious food by growing a pizza garden, packed with the tomatoes, onions, peppers, oregano, and basil that you need to make tasty home-grown pizza toppings. These vegetables all thrive during summer in the same conditions, and will grow well together in a large container or a section of a bed. If your kids don't like pizza, try other themed plantings that appeal to your children's food preferences – how about spring onions, oriental leaves, and mangetout peas for a stir-fry?

NEED TO KNOW

- Dress your kids in old clothes, and make sure they wear sturdy shoes.
- Supervise the use of sharp cutting tools at all times.
- As they tend to get muddier than grown-ups, keep kids off freshly manured soil. Always wash hands in soapy water after gardening.
- Show children which parts of each plant are good to eat; most crops are entirely edible, but tomato and potato foliage are toxic if eaten.

STARTING FROM SEED

Growing veg from seed is far cheaper than buying plants from a nursery and gives you the widest choice of varieties. The seeds of many crops can be sown directly into beds or containers outdoors, while others should be sown under cover and later moved to their final positions (*see pp.32–33*). Hardy crops can be sown outdoors as soon as the soil begins to warm (indicated by the sight of weeds); you can even begin a little earlier if you warm the soil by covering it with some horticultural fleece. Tender summer crops, such as runner beans, can be sown from late spring, once nights are no longer frosty.

SOWING INTO BEDS

There are few things more satisfying than watching plants grow to fill your vegetable bed. Sow when the soil is warm and moist, and there is little wind to send small seeds and packets flying. Water rows before sowing if the soil is dry. Sowing in orderly rows makes it easier to distinguish vegetable seedlings from weeds as they emerge.

PREPARING THE SOIL Create a fine breadcrumb-like texture on the surface of the soil by pulling a rake backwards and forwards to break down large lumps and remove stones. This allows you to draw out an even trench for sowing and permits the delicate roots and shoots of seedlings to push through the soil easily.

SOWING STEP BY STEP

1 Stretch a length of string across the bed between sticks or pegs to mark a straight line for your row.
2 Use the corner of a hoe or a trowel, to draw out a v-shaped trench along the line, at a depth of 1–5cm (½–2in), depending on the crop.
3 Sprinkle small seeds thinly along the base of the trench or drop in larger seeds at the recommended spacing.
4 Push soil back to cover the seeds using the back of a rake or your hands; add a label and water thoroughly.

HANDLING SEED

Some seed packets contain hundreds of seeds, while others hold as few as five. To avoid spilling them, tap the pack before opening to knock the seeds to the bottom. It is hard to sow small seeds, such as carrots and lettuce, evenly from a packet; instead, tip them into the palm of your hand and sow between thumb and forefinger. Larger seeds, from beetroot up to broad beans, are easy to pick up individually and place at their recommended spacing along a row. As soon as you have finished sowing each crop, return any unused seeds to the packet and fold over the top to seal.

NEED TO KNOW

- Label each row that you sow immediately after sowing.
- Add supports for climbing plants before or just after sowing.
- Weed around seedlings regularly to prevent competition and so give your crop the best start.

Tap seeds out of the packet into the palm of your hand so that you can easily control the number of seeds sown.

Seeds in containers are best sown at their final recommended spacing.

SOWING IN CONTAINERS

How you sow into large containers depends on the crop. The small seeds of plants such as cut-and-come-again salad leaves and early carrots can be scattered thinly across the surface and covered lightly with compost. For larger seeds, poke a hole in the compost to the recommended depth; sow two or three seeds into each hole and push compost over them with your hands. Water thoroughly.

THINNING

Seedlings grow rapidly and will soon begin to compete with their neighbours unless they are thinned to their recommended spacing. Start thinning as soon as the small plants are large enough to handle, by pinching them off at soil level or gently uprooting them. Thin gradually in case any seedlings are lost to pests. Firm the soil around remaining plants after thinning and remove any debris to avoid attracting pests. The thinnings from many vegetables, including lettuce, beetroot, and kale, are delicious added to salads.

TOP TIP COVER SEEDED BEDS WITH NETTING TO PREVENT CATS OR BIRDS SCRATCHING UP THE SOFT SOIL.

Thin in stages, starting with the weakest seedlings and any outliers from the row. After about two or three weeks, thin to the final spacing.

STARTING WITH PLANTS

Rather than sowing seeds directly into your garden beds, you can begin with young plants that you have bought or raised yourself under cover (*see pp.32–33*). This approach has many advantages: you don't need to worry about your seeds failing to germinate outdoors or your tender seedlings being eaten by hungry pests or damaged by bad weather; you'll also get a head start on the growing season. If you buy plants, however, be aware that the choice of varieties is limited compared to seeds.

CHOOSING HEALTHY PLANTS

Always look for strong, stocky plants with deep green leaves, and avoid pale, spindly specimens that display any yellowing or brown marks. Look at the base of the pot; if you see a dense mat of roots inside the pot or growing out of the drainage holes then the plant has been in there too long and you should choose another. Consider, too, where the plants are on sale; courgettes and chillies displayed outside on a cold spring day may already have suffered serious damage. The whole point of buying plants is to make life easy and get ahead, so don't start with stressed plants or risk introducing disease to your veg patch.

At the garden centre, look for lush foliage; avoid plants with wilting leaves and very dry compost as this may indicate neglect.

Reuse old plastic bottles as cloches to protect young plants.

HARDENING OFF

Young plants that have been raised in a nursery or under cover at home will be accustomed to warm conditions and liable to suffer a damaging shock when planted outdoors. This can be avoided by gradually acclimatizing plants to cooler conditions over a week or two before planting out. To "harden off" plants in this way, simply move them in their pots outside for a longer period each day, bringing them under cover again at night. Alternatively, cover them with cloches for a week or two once they are planted out. Hardening off is particularly important for tender summer vegetables, which are easily damaged by cold and windy weather.

PLANTING OUT

Before planting, make sure the area is weed free. Level the soil and water the plants in their pots. This procedure is the same whether you are planting into the ground or into containers filled with multi-purpose compost.

PLANTING STEP-BY-STEP

1 Remove the plant from its pot by gently pulling its stem or placing your hand over the compost and turning it upside down. Press the bases of modules to push plants out. Retain as much compost around roots as you can.

2 Dig a hole to the same depth as the pot with a trowel or your hand. Place the rootball into the centre of the hole with the plant upright.

3 Push soil around the roots to fill in any gaps and gently firm around the plant with your hands. The soil should reach the same level on the plant's stem as the soil did in the pot.

4 Water the plant to help settle the soil around the roots. Repeat the process for subsequent plants, making sure to allow the recommended space between neighbours.

TOP TIP PROTECT NEWLY PLANTED VEGETABLES WITH SLUG TRAPS OR NETTING TO PREVENT PESTS FEASTING ON THEIR TENDER LEAVES.

PLANTS BY POST

If your local nursery doesn't have the variety you need, try an online or mail order supplier. Before placing an order, check whether plants will be delivered as small plug plants (which may need to be grown on in pots before planting out) or larger plants in 9cm (3½in) pots. If you are ordering more than one crop, find out when each will be delivered; they may be dispatched at different times. Make sure you're home when live plants are delivered; they need to be opened and planted as soon as possible.

Plants sent by reputable nurseries are well packaged and healthy.

PLANTING OUT: TIMETABLE

Early to mid-spring
Beetroot, broad beans, lettuces, onions, peas.

Late spring to early summer
Aubergines, cabbages, chillies, courgettes, French and runner beans, leeks, peppers, squashes, sweetcorn, tomatoes.

Mid-summer to early autumn
Kale, oriental greens, spring cabbage, sprouting broccoli, winter lettuces.

STARTING OFF UNDER COVER

Raising plants under cover on a sunny windowsill may seem like more hassle than sowing seed directly outdoors, but it has many advantages. Most importantly, it gives you some control over temperature, which produces more reliable germination and allows you to extend the effective growing season for slow-maturing veg and for tender crops that can't be sown outside until early summer. As your plot develops and your gardening ambitions expand, it may be worth investing in a mini greenhouse or glasshouse to provide extra protected growing space outdoors.

Put pots on drip trays to prevent water from damaging the windowsill.

WINDOWSILL NURSERY

Any bright windowsill in a warm room will provide good conditions for seeds sown in pots or module trays to germinate. Once the seeds are sown, place each pot or tray in a clear plastic bag or into in a simple propagator with a clear plastic lid; this will help to retain moisture and warmth. Remove the bags or lid when seedlings appear and then turn the pots or trays daily to stop seedlings being drawn in one direction. Try not to shut plants behind curtains at night, where the air trapped against the window can become cold. Avoid sowing too early, because plants kept waiting on a windowsill for too long before planting outdoors can become leggy and pale.

HEATED PROPAGATORS

Any added warmth will help accelerate germination, but consistent heat from underneath the pot or seed tray – such as that provided by a dedicated heated propagator – is the ideal way to nudge seeds into life. Heated propagators give the compost temperature a useful boost in a cool porch or conservatory, and if used in a warmer room, can help push growing temperatures up to around 21°C (70°F), which is sufficient for the reliable germination of heat-loving crops like chillies and aubergines.

Heat from the base also makes germination more predictable; it is then far easier to time sowings accurately to give you plugs to transplant exactly when you need them. Choose one of the many reasonably priced electric propagators that are designed to fit neatly on windowsills.

TOP TIP TRANSPLANT SEEDLINGS GROWING ON A WINDOWSILL INTO PROGRESSIVELY LARGER POTS TO AVOID CHECKING THEIR GROWTH, PARTICULARLY IF COLD WEATHER CAUSES A DELAY IN TRANSPLANTING THEM OUTDOORS.

Propagators are inexpensive to run. Choose one with a thermostatic control to enable you to set the temperature accurately.

SOWING INTO MODULES

Module trays are divided into separate growing compartments for individual plants and are the easiest way to raise your own high-quality "plug" plants. They are worth buying because they remove all worries about spacing seeds correctly when sowing and eliminate competition between seedlings. Using these flexible trays makes it easy to transplant seedlings without disturbing their delicate roots with fiddly pricking out. Most seed trays are made from recycled plastic and have drainage channels to ensure that plants don't sit in water.

SOWING STEP-BY-STEP

1 Fill all the modules by scooping multi-purpose compost over the tray, gently pushing it down into each module with your fingers, then topping up with more compost.
2 Use a pencil or your finger to make a hole in the centre of each module to the recommended planting depth.
3 Sow a single seed into each module, cover with compost, label, and water thoroughly before placing on a windowsill.

PROTECTING PLANTS OUTDOORS

If you don't have suitable space indoors, consider providing cover in your garden. A simple option is a mini-greenhouse, a rack for pots that can be enclosed under a clear plastic zip-up cover. More robust versions have a wood or metal frame and polycarbonate or glass panels. These shelters are designed to stand against a wall or fence and are ideal for protecting seedlings from the cold, but need careful ventilation to minimize damaging temperature fluctuations.

GLASSHOUSES These more durable and expensive structures allow you to grow seedlings on staging alongside tender vegetables, such as tomatoes and peppers, in colder regions where they might otherwise fail to ripen outdoors.

Ventilate glasshouses during the day to control the airflow and temperature.

TOP TIP CHECK YOUR LOCAL FREE AD LISTINGS: GLASSHOUSES ARE SOMETIMES OFFERED FOR FREE TO ANYONE WILLING TO DISMANTLE, TRANSPORT, AND REASSEMBLE THEM.

NEED TO KNOW

- Once plants have germinated, they require as much light as possible, but may need to be shaded from intense midday sun on a south-facing windowsill.
- Good air circulation is important to prevent fungal diseases, such as damping off. Always remove covers from propagators and pots after germination and leave sufficient space between larger plants.
- Check the moisture levels of compost daily and water all plants growing under cover regularly.
- Harden off all indoor-raised plants for about two weeks to toughen them up before planting them outside.

FEEDING

To produce plentiful, healthy, and tasty crops, your veg plants need ready access to nutrients throughout their life cycles. The three main nutrients required to fuel growth are nitrogen, phosphorus, and potassium, but plants also need tiny quantities of essential trace elements, such as iron and magnesium. You can buy commercial fertilizers that deliver different amounts and types of nutrients to your plants. However, this can be costly and is a poor substitute for improving your soil by adding generous quantities of organic matter every year.

FEED THE SOIL

Seasoned vegetable gardeners swear by the maxim "feed the soil and your plants will feed you". That's because veg plants grown in fertile soil, enriched every year with well-rotted compost or manure (*see pp.46–47*), can obtain all the nutrients they need with little or no need for fertilizers.

The regular addition of organic matter provides food for earthworms and beneficial soil microorganisms, which gradually break it down further into humus. In contrast to the single rapid burst of nutrients delivered by most fertilizers, this natural process of decomposition gradually releases vital nutrients from the organic material in forms that are accessible to plant roots. Feeding the soil in this way gives it a crumbly texture and a rich colour, and increases its ability to hold moisture. It produces steady growth that results in sturdy plants that are less vulnerable to pests and diseases. What's more, this approach ensures that the harvest from your plot improves year on year as your soil becomes progressively more fertile.

TOP TIP USE FERTILIZERS IN A NEW VEGETABLE PLOT TO SUPPLY PLANTS WITH KEY NUTRIENTS WHILE YOU ADD ORGANIC MATTER TO THE BEDS EACH YEAR TO IMPROVE THE SOIL.

Applying a compost mulch to your soil every year will improve its structure, buffer pH, and increase the population of beneficial organisms.

CONTAINER NEEDS

Vegetables grown in containers can only spread their roots through a limited amount of compost, which typically holds enough nutrients to last for about six to eight weeks. After this point, pot-grown plants rely on you for food, so it's vital to apply the correct fertilizer regularly to achieve a good crop. This is especially true for "hungry" plants, such as tomatoes and courgettes, which need a constant supply of nutrients to fuel their rapid growth and abundant fruit production. Liquid feeds are easy to apply to containers during watering; choose a balanced fertilizer for leafy crops and a high-potash feed to produce good yields from fruiting vegetables.

Tomatoes grown in containers will need feeding with a fertilizer that is rich in potash.

TYPES OF FERTILIZER

Commercial fertilizers are used to feed plants directly, either where there is a shortage of nutrients in the soil, or to give plants a boost during a particular stage of growth. A huge range of both organic and inorganic fertilizers (*see right*) is available in granular or liquid forms. Apply granular fertilizers to the soil, either as a base dressing raked into the soil before sowing or planting, or as a top dressing sprinkled on to the soil surface around plants. Liquid feeds are usually diluted with water and applied using a watering can. It is important to keep fertilizers off foliage, because they can scorch leaves. Always follow the manufacturer's instructions and check the suitability of each product for use on edible crops.

Liquid fertilizers are mixed into the plant's water supply.

NEED TO KNOW

The proportions of the three main plant nutrients in a fertilizer are shown on its label as the N:P:K ratio. All three numbers are the same in a balanced fertilizer.

- **Nitrogen (N)** promotes the growth of leaves.
- **Phosphate (P)** promotes root and shoot growth.
- **Potassium (K)** (also called potash) promotes flowering and fruiting.

ORGANIC VS INORGANIC

Consider which fertilizers you are happy to use on your home-grown crops. Organic fertilizers are from plant or animal sources: they include products such as seaweed extract, comfrey concentrate, blood, fish, and bonemeal, and pelleted poultry manure, which are slower acting but longer lasting than inorganic fertilizers.

Often cheaper, inorganic fertilizers, such as sulphate of potash and sulphate of ammonia, are produced synthetically or mined. They are concentrated and fast-acting, delivering nutrients to roots in the forms that plants can use immediately.

General-purpose fertilizer is available in convenient granular form.

WATERING

Water is essential for every stage of plant growth, which makes watering among the most important jobs in the garden. Location plays a key role in determining how much watering a plant requires. Crops in warm regions will be thirstier than those in cool, wet climates, and plants in containers and free-draining raised beds will need more watering than those in open soil. Any vegetables grown under cover will receive no rainfall and be totally reliant on watering. The regular addition of organic matter to soil increases its ability to hold moisture and reduces the need for watering.

Water early and late in the day to minimize evaporation and maximize wetting.

WHEN TO WATER

Seeds need water to germinate, as do young plants to establish, so you should always water the soil after sowing seeds or transplanting seedlings. After that, veg growing in good, fertile soil can often manage without additional watering, but a few well-timed soaks will increase yields considerably. Concentrate on watering fruiting and pod crops during flowering and root crops in their later stages of growth (to help swell roots); leafy crops need only occasional heavy waterings throughout their growth.

Be responsive to conditions: all plants need extra watering in spells of hot, dry weather, but you should also reduce watering when it is cool and cloudy, because overwatering can be just as damaging as underwatering.

HOW TO WATER

It takes a surprisingly large amount of water to permeate through dry soil to a depth that is useful for roots. Only wetting the top layer of the soil encourages shallow rooting and prevents plants sinking roots deep into the soil where more moisture is naturally available. This means it is much more beneficial to give plants growing in the soil an occasional good, long soak than a daily trickle: aim to apply the contents of two large 10-litre (2.5-gallon) watering cans per square metre (square yard) of soil. Use a watering can with a rose, or a hose with a spray attachment, to water gently with fine droplets. This prevents damage to the soil structure and stops gushes of water washing away fine seeds.

Water all around the base of a plant to promote even root development.

CONTAINERS

Compost dries out rapidly, even in large pots, so container plants need frequent watering: in summer, this can mean once or even twice a day. Established plants will act as umbrellas, preventing rainwater reaching the compost in the pot, so be sure to water such plants beneath their foliage.

Pots standing in trays can be watered from the base, but never leave a pot standing in water. If you're going away from home, set up an irrigation system using seep or drip hoses connected to a mains tap or water butt, and add a timer to deliver the correct amount of water to your plants. Such systems are readily available from garden centres.

Drip irrigation systems deliver water consistently and economically.

NEED TO KNOW

- Avoid heavy watering of fruiting crops like tomatoes and chillies during ripening; too much water can reduce their flavour.
- Water consistently over time, because irregular watering can cause roots and fruits to split, and leafy crops to bolt.
- Plants stressed through lack of water are more vulnerable to disease.
- Avoid overwatering: standing water will damage a plant's roots.

TOP TIP WATER THE SOIL AT THE BASE OF A PLANT, AVOIDING THE LEAVES. WET LEAVES MAY "BURN" IN SUNLIGHT AND BECOME MORE SUSCEPTIBLE TO FUNGAL INFECTION.

USING RAINWATER AND GREY WATER

Plants don't need clean tap water to thrive. Making use of other water sources is good for the environment and will save you money if your supply is metered. Collect rainwater as it runs off the roof of your house or shed by diverting a down pipe into a water butt (kits are available from garden centres). Saved rainwater can be used to water everything except seeds and seedlings, which may be affected by microorganisms in the water.

Grey water is domestic water from washing dishes and bathing. It can be used to water established plants that will be cooked before eating. Diluted washing detergents won't harm plants, but don't use water containing strong cleaning products. Never store grey water for more than 24 hours before use to stop bacteria proliferating.

Fit a lid to your butt to prevent animals gaining access.

PROTECTING YOUR CROPS

Raise plants well in healthy soil and they will be strong enough to resist and outgrow the majority of pests and diseases. Good garden hygiene also helps to achieve a healthy crop, so remove any weeds and dead material that could harbour problems. Encouraging beneficial wildlife into your garden can keep pest numbers under control without the use of chemical sprays, but this natural balance will never remove pests completely. Luckily, physical barriers are an effective way to keep plants and pests apart, while traps and other distractions can also be employed to keep crops safe.

STOPPING SLUGS AND SNAILS

Whether you grow veg in containers or in open soil, slugs and snails are likely to be your worst enemy. They are voracious night-time feeders, particularly in wet weather, and can wipe out rows of seedlings overnight. It's not possible to eradicate them, but numbers can be reduced by regular torchlight hunts and ensuring there are no dark, damp daytime hiding places close to your plot. Create traps in vegetable beds by placing scooped out grapefruit halves or sinking jars half full of beer into the soil, leaving overnight, and disposing of captives in the morning. Copper tape around pots can repel slugs (*see p.19*), as can prickly obstacles, such as crushed egg shells or pine needles, around vulnerable plants.

Some chemical slug pellets can be used around edible plants, but they are often considered a risk to children, pets, and wildlife. Biological control with preparations of nematodes (tiny worms that attack slugs and snails) is possible but relatively expensive and is usually only worthwhile in a larger garden.

A barrier made from crushed egg shells may help to keep slugs off your salad crops.

TOP TIP SNAILS OFTEN GATHER TO HIBERNATE IN SHELTERED PLACES OVER WINTER. SEARCH FOR THEM IN LATE WINTER OR EARLY SPRING AND REMOVE ANY THAT YOU FIND TO GIVE SPRING SEEDLINGS THE BEST CHANCE.

Slugs love citrus, so a half-grapefruit or orange makes an effective trap.

Netting is an effective barrier. You can water your plants through the netting.

PHYSICAL BARRIERS

Simple barriers often are the best way to keep pests away from crops. Protect peas and brassicas from wood pigeons by growing them under netting; this needs to be supported on a substantial frame to allow space for tall plants like sprouting broccoli. Rabbits can only be excluded by enclosing the growing area inside a fence at least 1m (3ft) high and buried to a depth of 30cm (1ft). Crops that need protection from small insect pests are best grown under tunnels of horticultural fleece or fine insect mesh, supported by curved lengths of wire or blue alkathene pipe. When covered in fleece, these tunnels have the benefit of protecting plants from cold weather and extending the growing season.

COMPANION PLANTING AND SACRIFICIAL CROPS

Some plants help to deter insect pests or attract predators that feed on pests of veg crops. Try planting French marigold (*Tagetes*), onions, or chives, among carrots to distract carrot flies from their scent. Poached egg plant (*Limnanthes*) planted close to broad beans brings in hoverflies, which feed on aphids, while later-flowering fennel does the same for runner beans. Growing flowering plants among vegetables also attracts pollinating insects to help increase yields.

Sometimes specific plants are used as sacrificial crops to draw pests away from vegetables. Nasturtiums are often used in this way, because they are irresistible to aphids.

Nasturtiums are cheery plants that are easy to grow and attract blackfly away from the shoot tips of broad beans.

Hoverflies, ladybirds, and lacewings are your allies in the war against aphids.

ATTRACTING BENEFICIAL WILDLIFE

Birds, hedgehogs, frogs, and a whole range of insects and other invertebrates prey on common garden pests. It's easy to take a few simple steps to encourage a more diverse ecosystem in your garden. Feed birds, put up nest boxes if you have space, and provide them with fresh water year-round and they will visit your garden regularly, especially if there are trees and hedges for cover.

Keep your veg plot tidy, but leave one corner a little wilder, with fallen leaves for hedgehogs, a small log pile for invertebrates, and maybe even a pool made in an old pot to attract frogs and toads. Plant a tree or hedge if you are able to: this greatly expands the range of habitats available to insects, birds, and mammals.

NEED TO KNOW

- If you choose to use chemical controls, always check they are approved for use on edible crops.
- Follow the manufacturer's instructions precisely.
- Never spray insecticides on flowering plants and always apply late in the evening to avoid killing beneficial insects, such as bees.

AFTER THE HARVEST

A bumper crop in early autumn is always a great reward, but your work in the garden doesn't stop there. Many vegetables yield far more than you can eat fresh, so you'll need to prepare and store your produce to see you through the winter. Autumn is also the time to tidy your plot and recycle any plant waste on a compost heap to produce free soil improver for next year. Make the effort to record your successes and failures at the end of the season: this will help you identify the crops and practices that work best for you, and enable you to grow into an expert gardener over the years.

STORING YOUR HARVEST

Many crops can be dried, frozen, or preserved. Those that naturally keep well include onions, shallots, and garlic, which just need to be laid in the sun to dry their foliage, before being hung in bunches somewhere warm and dry. Borlotti and haricot beans that have matured and been allowed to dry fully in their pods are easily stored in airtight jars in the kitchen, ready for cooking.

Once their skins have toughened in the sun, winter squashes will also keep for months indoors in a cool, frost-free room, where they can be arranged to make a colourful display. Root crops, such as beetroot, carrots, and turnips, can be lifted from the ground, placed in boxes packed between layers of sand or sieved sandy soil, and stored in a frost-free shed or garage.

Beans, peas, spinach, chillies, and herbs are all quick and easy to freeze, while beetroot, tomatoes, and runner beans can be cooked with sugar, vinegar, and spices to make richly flavoured chutneys.

Let winter squash mature and dry out on the plant; remove it for storage before the first frost.

Stored correctly, carrots will retain their nutrients for over three months.

PREPARING FOR WINTER

After you have harvested most of your crops, it's time to start getting your growing space ready for winter. Pull up spent plants except for peas and beans, which should be cut at soil level to leave their nitrogen-rich roots in the soil. Compost healthy plant material (*see pp.46–47*), but put anything showing signs of disease into the domestic waste bin to prevent problems spreading. Clear out the remnants of last year's compost, spreading it on your beds to enrich the soil (*see pp.20–21*), and making room for new material to be composted.

CLEAN UP Giving your garden a thorough tidy not only makes it look neater but reduces opportunities for pests and diseases to overwinter. Remove weeds, sweep up debris between pots, and clear away supporting canes, netting, and fleece. Clean your seed trays and pots so that they are ready for the next growing season, and disinfect the interior and surfaces in your greenhouse, if you have one. Give your garden tools some attention too: wash and dry metal parts before rubbing them with an oily rag to deter rust. Treat wooden handles with linseed oil, which stops the wood from drying out and becoming brittle.

COVER UP Even though growth slows down greatly over the winter, it pays to not leave your soil bare over this period. Nutrients are liable to be leached out by winter rains, and bare soils are more prone to becoming waterlogged, killing beneficial bugs and microbes. Instead, apply a thick autumn mulch of well-rotted compost over the soil surface in autumn, or sow a green manure or cover crop, such as grazing rye or winter tares (vetch), in early autumn, to protect the soil from winter weather. These plants can be dug into the soil in spring to add organic matter and nutrients for the next crop.

Winter tares is a green manure that will provide cover and enrich your soil.

Clean your greenhouse with a disinfectant solution to prevent disease.

Making notes will help you grow into a successful gardener.

KEEP A JOURNAL

Gardeners have always kept diaries or journals to help them learn from experience with the passage of each year. It's worth keeping notes yourself to record the varieties of each type of vegetable that you have grown, the dates that they were sown, and the timing and quantity of the harvest of each crop. Paste seed packets and photos into your notebook and make sketches of successful arrangements of crops or supports. All this will serve as a useful reminder of what is worth growing again when you come to buy seeds the next year.

Knowing exactly how long each crop is in the ground also makes it possible for you to plan a succession of crops to follow on from one another throughout the year (*see p.22*) and so optimize every inch of growing space. Record any problems you've encountered with pests or diseases too, along with the dates of any late frosts, to help you develop a unique way of gardening that works for your plot. Online garden journal apps are a modern twist on this old idea and make it easy to add photos and have fun sharing gardening tips and successes with friends.

YOUR SECOND SEASON

Planning your second and subsequent seasons of growing veg is an opportunity to bring more land into use to increase yield, to introduce new growing techniques, and to try tasty and exciting new crops. Even if you choose to stick to what you know and like, the vagaries of the weather, pests, and diseases mean that every year is different and presents new challenges. Use your notes from the previous years to work out what flourishes, what fails, and why. This information is also key to finding the best times to sow and plant, and avoiding repeated mistakes and wasteful gluts.

Consider if hardware, such as irrigation systems, would benefit your garden.

PLAN FOR SUCCESS

When there is less to do in the garden over winter, carry out some research on how to get the best from your plot. Consider, for example, how you could be more productive by having follow-on crops ready to plant as soon as one crop is harvested; plan some plant combinations for intercropping, or think about dead spaces in the garden that could support climbing veg to make use of every available space. Now that you are familiar with local conditions, scour seed catalogues for varieties that suit them best. Try dwarf varieties in a windy garden, cold-tolerant or fast-maturing selections in more northerly locations, or vegetables bred with resistance to a troublesome pest or disease.

TRY SOMETHING NEW

Assess your experiences with each crop in your garden. If a plant didn't do well first time round, consider why. Was it sown too early or too late? Did it get damaged by frost or spoiled by a pest? If you have a sound plan for a remedy, give that crop another try. If you don't, then avoid repeating the experience: instead, try a similar crop that may be easier or more forgiving to grow; for example, swap kale for cabbage, Swiss chard for spinach, or chives for spring onions. Changes can also be made where the weather has affected a crop's performance. So try growing dwarf beans if your climbing beans were damaged by high winds, or switch to faster-ripening cherry tomatoes if last year's beefsteak variety remained stubbornly green. Equally, don't feel compelled to grow something that produced a great harvest if you didn't enjoy eating it.

Chard is easier to grow and more vigorous than spinach; however, there are no maintenance-free plants!

MANAGE QUANTITIES

In your first year of gardening, you'll probably have a glut of some veg and a tiny crop of others. For subsequent years, consider how much space you really want to give each crop in your garden. Some, such as potatoes, squashes, and cabbages, are large plants, and growing just one or two fewer could free up valuable space to grow more of the crops you love; similarly, perhaps you would prefer a second, late row of peas to an overabundance of runner beans. Make small successional sowings of root crops and salads for continuous harvests rather than a single end-of-season glut.

Cabbages are a great crop if you have the space to accommodate them.

NEED TO KNOW

- Always clean your pots, seed trays, and tools after use to prevent transmission of disease.
- Always sow seeds and grow container plants in fresh compost; this will help to reduce problems with pests and diseases.
- Remember to plan around any trips or holidays so there aren't lots of seedlings to water or crops to harvest when you're away.
- Time spent improving your soil is never wasted. It is the basis of all successful gardens.
- Fresh seed has the best germination rates, especially for carrots and parsnips.

BE REALISTIC

When you realize that you can grow tasty vegetables successfully, you may be tempted to ramp up your production in the following year. Be careful, however, not to take on so much that you struggle with the commitment in time and energy. Increase your growing space gradually and don't be afraid to rein things in: gardening isn't meant to be stressful. Be realistic, too, about working within the constraints of the local climate. If heat-loving crops like aubergines and chillies perform poorly outdoors, then try them on a windowsill instead; and if leafy crops wilt in hot, dry weather, find them more shade or don't grow them over the summer.

TOP TIP IF YOU ARE GROWING VEG TO SAVE MONEY, THEN YOU WILL – IN THE LONG RUN. THE FIRST COUPLE OF YEARS WILL INEVITABLY COST YOU MORE BECAUSE OF INVESTMENT IN YOUR SOIL AND TOOLS, AND A HIGHER RATE OF FAILURE AS YOU LEARN.

A neat, small raised bed packed with your favourite ingredients will give more satisfaction than a large, difficult, messy plot.

EXTENDING THE SEASON

A bumper summer harvest of homegrown veg is a thrill, but with a little forethought you can have crops available to pick throughout the year so that they become an everyday pleasure. Selecting varieties carefully and sowing small amounts at regular intervals will give you an abundance of vegetables that mature in succession so that they can be enjoyed at their peak over a long period. The growing season can also be extended significantly by protecting spring sowings, and vegetables maturing in autumn, from cold weather using cloches or coverings of horticultural fleece.

Radishes can be sown from seed and harvested just 3–4 weeks later.

SUCCESSIONAL SOWING

Rather than sowing all of your seeds at once in spring and producing a glut of veg in summer, make small sowings at intervals throughout spring and summer. Such successional sowing is the best way to produce a continuous supply of fast-growing crops, such as salad leaves, radishes, and baby beetroot. Sow these vegetables every 2–3 weeks, or when the previous sowing starts to poke through the soil.

The growing seasons for slower-growing crops, such as peas, dwarf French beans, spinach, and kale can also be extended by making several sowings through spring and summer.

MIX VARIETIES

Different varieties of vegetables often mature at different rates. Use this to your advantage by sowing fast- and slow-growing varieties to harvest one after the other. As a rule, any variety described as "early" tends to mature faster than the rest, which are often known as "maincrops"; this is the case for potatoes, peas, and carrots. Dwarf and small-fruited varieties are also usually quicker to crop because they have less growing to do, so cherry tomatoes ripen earlier than large-fruited varieties, especially those growing on bushes rather than tall cordon plants; and dwarf French beans can be harvested a few weeks ahead of climbing varieties.

Plenty of crops can also be harvested at differing stages of maturity: young beetroots, for example, can be used for salads, older ones for stews and soups, effectively extending their season.

Early carrot varieties develop rapidly from spring sowings to produce sweet young roots ready to pull from the soil by midsummer.

KEEP OUT THE COLD

If you live in a cooler climate, you can extend the growing season by up to a month in both spring and autumn simply by covering plants to protect them from the cold. This technique allows spring sowings to be made earlier and shelters late crops as the weather turns cold in autumn.

Cloches are usually shelters for individual plants. They were traditionally made from glass but are now available in hard plastic. Effective and low-cost cloches can be made by cutting the bases off large, clear plastic bottles.

If you want to protect rows of plants, a better alternative is horticultural fleece. This is a light, flexible fabric that can be laid over the soil like a blanket and secured at the edges. Its advantages are that it stretches as plants grow (although it's best to leave a little slack), allows rain through, and provides a useful barrier to pests. Tunnels covered with fleece or clear polythene are the best way to protect taller crops. Hoops of sturdy wire or alkathene pipe (which you can buy from a plumber's merchant) can be cut to fit the bed's dimensions and will provide pliable supports for the fleece or plastic covering.

Cloches made from water bottles are ideal shelters for newly planted seedlings and help keep snails away.

NEED TO KNOW

- Weeds also thrive under crop covers, so hoe or hand weed regularly.
- Make sure to regularly water plants growing under cloches or plastic tunnels.
- Remove or prop open crop coverings on warm days to allow air to circulate and help prevent fungal diseases.
- Anchor cloches and fleece securely to prevent them blowing away.

Chillies will continue to ripen in autumn given a warm, bright spot indoors.

MOVE INDOORS

To keep container plants cropping for longer, bring them indoors as the weather gets cooler. Place them on a windowsill or in a warm, sunny porch or conservatory. In this way you can ripen your last chillies, peppers, and aubergines when the autumn weather makes it impossible for these tender plants to survive outside. Herbs can also be potted up in late summer to be brought indoors for picking into autumn and even winter. Some smaller plants can even be lifted from the garden and potted indoors.

TOP TIP TRY LIFTING AND POTTING UP PERENNIALS LIKE MINT, CHIVES, AND MARJORAM, OR MAKE LATE SOWINGS OF PARSLEY, BASIL, AND CORIANDER IN POTS FOR THE KITCHEN WINDOWSILL.

Windowsill herb crops can be picked when outdoor plants have died down.

COMPOSTING

A supply of well-rotted compost is essential for success with veg. You can make your own compost for free in as little as six months by recycling your garden and kitchen waste on a compost heap. This will turn the remnants of harvested plants, vegetable peelings, grass clippings, and prunings into valuable crumbly compost to improve your soil. Compost bins are available to buy in a range of designs, but building your own costs less and lets you make optimal use of the space available. Make your bin as large as possible, because the bigger the heap, the faster waste breaks down.

BUILDING A COMPOST BIN

Wooden pallets are an ideal starting point for making your compost bin. In this design, three standard-size pallets (1.2m/4ft x 1m/3ft) are screwed together to create the sides of the bin, while a fourth pallet at the front can be opened to allow easy access for turning and emptying. This heap is a good size for a large, busy garden, but smaller bins of a similar design are easy to make by knocking four posts into the ground and using chicken wire lined with thick cardboard to form the sides.

CONSTRUCTION STEP-BY-STEP

1 On level ground, position three pallets to form the static sides of the bin. Make sure the fourth pallet fits snugly into the gap at the front, where it will form the door.

2 Once the pallets are neatly positioned, secure the three side pallets together at the corners by driving two long screws into the thick blocks of wood at the top and bottom of each pallet.

3 Line the insides of all four pallets with chicken wire, by hammering in staple nails, to hold all of the material added to the heap in place.

4 Lift the fourth pallet into place to fill the gap at the front. Make a hinge by loosely tying it to the adjoining pallet with wire down one side. Secure the opening side with a single wire tie.

SITING AND FILLING

If possible, site your bin on bare soil, where worms and other decomposing organisms have easy access and can get to work faster. It is tempting to hide the compost bin away in a dark corner, but a sunny site is better, as warmth speeds up the composting process.

Fill the bin with a mixture of about 50:50 leafy green and twiggy brown waste: too much green and the heap will become slimy and smelly; too much brown and it will be too dry to decompose. The larger the heap, the better the composting process works.

Compost bins can be hidden effectively among cheery blooms.

NEED TO KNOW

- Add more carbon-rich brown waste (including cardboard, wood chippings, and paper) if the heap starts to become a bit slimy or smelly.
- Don't put perennial weeds or annual weeds that have set seed on to your heap because they will pop up again when the compost is spread.
- Don't add meat or cooked food waste because it attracts vermin.
- Don't add diseased plant material to the compost: dispose of it in the domestic waste bin instead.

Covering the compost heap accelerates the process of decomposition.

KEEP A LID ON

Once you begin adding waste to your bin it is important to cover the top, primarily to stop the contents getting soaked by heavy rain. The cover can be made from an old piece of carpet, thick cardboard, or black plastic, weighed down to prevent it blowing away. Covers help accelerate the composting process by insulating the bin and allowing the heat generated by decomposition to build up and kill any weed seeds and diseases. They also protect the compost from weed seeds that are scattered on the wind or by birds.

TURNING THE HEAP

The process of decomposition doesn't happen evenly throughout a compost heap. The centre of the heap stays moist and becomes warm as it breaks down quickly, while the edges tend to dry out and decompose slowly. Moving the material around helps it to decompose evenly, so you should turn the compost at least once during the process, but it is easier and more effective if done monthly. Use a garden fork to pull material from the sides of the heap on to the top, then push material from the centre out to the edges and mix. This is a good time to add water if the heap looks dry. Always chop up woody material, including cabbage and broccoli stems, and chip branches before composting.

TOP TIP IT'S EASIEST TO HAVE TWO COMPOST HEAPS; ONE THAT YOU'RE ACTIVELY FILLING AND ANOTHER WHERE THE FINISHED COMPOST CAN BE KEPT AND USED AS REQUIRED.

Turning aerates the compost, allowing oxygen to reach decomposing bacteria deep in the heap.

MAKE MORE SPACE FOR VEG

Once you've tasted home-grown veg and appreciated the benefits of producing your own food, you'll probably want to expand your operation. There's no need to move house – a little lateral thinking could free up sufficient additional space to let you get even more creative in the garden.

First, take advantage of any parts of your plot that are currently underused – a wall or fence could support productive climbers. Then, consider your priorities: do you really need all that lawn; could you park the car further up the drive, does anyone use the patio for anything except drying clothes?

CONSIDER A RAISED BED

Raised beds are the best way to quickly extend growing space (*see pp.16–17*). They take a little investment in time and effort to build, but they are easy to maintain thereafter, and ideal for growing vegetables because you can reach easily over the soil.

Choose a site that receives at least four hours of sunlight per day, and avoid surfaces that are prone to rotting (such as wooden decking). Otherwise, raised beds can be built on any underused patch of land, such as a neglected paved patio, or an area of hardstanding, such as the end of a drive or a concrete base where a greenhouse once stood. Tired patches of lawn can also be good sites: there's no need to lift turf, just place the frame on the lawn, cover the grass inside with thick cardboard, then spread a mixture of soil and compost on top to fill the bed.

Raised beds built on hard surfaces should be at least 30cm (12in) deep to provide plants, especially root crops, with space to extend their roots.

Climbing beans produce a colourful display of flowers and pods.

EDIBLE BEDDING

Instead of buying summer bedding plants to fill your pots and borders every year, capitalize on the ornamental qualities of vegetable plants and herbs. Many tender vegetable crops thrive under the same conditions – and grow over the same period – as traditional half-hardy bedding, so it should be a straightforward swap. Why not flank your front door with bold pots of bush tomatoes, chillies, or aubergines, adorn the decking with colourful varieties of dwarf French beans, or grow ornamental squashes instead of sweet peas?

MIX WITH ORNAMENTALS

Many gardeners have discovered that small gaps between their ornamental plants can be usefully filled with edibles such as salad leaves. However, the possibilities don't end there because many veg plants are decorative in their own right and can add a creative twist to a classic ornamental planting. Try handsome cabbages and kale for autumn and winter structure, statuesque sweetcorn and climbing beans to add height to the backs of borders, or line paths with chives, thyme, or basil. For your veg to succeed in ornamental beds, you'll need to improve the soil with a thick mulch of compost each year, usually applied in autumn.

Front gardens are curiously underused for edible crops, especially when they are south-facing and receive more sunlight than the back. Yet another option is to combine ornamentals with veg plants grown in pots: pot marigolds, for example, pop out against the bold leaves of kohlrabi, and blue lobelias look particularly striking when planted alongside bright green lettuces.

TOP TIP PLANT CROPS THAT LOOK AS GOOD AS THEY TASTE. START WITH BORLOTTI BEANS, GLOBE ARTICHOKES, SQUASHES, AND BEETROOT.

This cottage garden looks the part with kale, lettuce, aubergine, and tomato planted alongside foxgloves and hostas.

GO OUT OF BOUNDS

Over time, your ambitions may exceed the capacity of your garden or terrace. It's then time to explore other options to find growing space. Renting a local allotment is the obvious place to start, but demand is often high, so be prepared to add your name to a long waiting list. At the same time, put the word out that you are keen to grow your own veg and a friend or relative may ask for help with their garden in return for a share of the harvest.

The gardening "grapevine" has been supplemented by social media groups and garden sharing apps that can put you in touch with others who have land to share; only proceed with these types of arrangements if you are comfortable with their terms and get on well with the garden's owner. Many community gardening projects also provide space for individuals to grow food, particularly in cities, and can be a great place to meet like-minded people.

Community gardens can offer ample space to grow veg as well as a place to gain advice from experienced gardeners.

Salad seeds are sometimes sold as mixes, giving you a variety of leaves from a single seed packet. Here the planting includes pot marigolds – companion plants that attract beneficial insects like hoverflies, which prey on aphids. Their colourful petals are also edible and look beautiful in salads.

SALADS AND HERBS

Growing your own means that no salad need ever be boring. There is a staggering range of juicy, crisp, sweet, peppery, or bitter leaves that you can produce quickly, cheaply, and repeatedly in the smallest of plots.

FRESH LEAVES YEAR-ROUND

Quick cropping and easy to grow, salads and annual herbs can produce a continuous supply of leaves throughout the year when sown successionally in small quantities every two or three weeks. Try sowing an array of lettuces, peppery rocket, basil, coriander, and salad mixes outdoors from early spring; and if you like punchy flavours, add the acidic tang of sorrel and intensely savoury leaf celery. For fresh leaves through autumn and winter, make late summer sowings of winter lettuce, radicchio, and endive to cover with cloches as the weather turns cold. Alternatively, shift production on to your windowsill for winter harvests of baby leaves and pea shoots.

COMPACT CROPS

Salads are one of the most productive crops for a small space. Shallow roots allow them to flourish in containers, which can be packed onto windowsills, balconies, or patios. Cut-and-come-again crops can produce two or three harvests of baby leaves from a small patch of compost, and are brilliantly suited to container cultivation. Rapid growth makes salads perfect for intercropping and filling any unexpected gaps in vegetable beds, which means there is always room to squeeze more in. Herbs will also thrive in pots, and can add a decorative flourish in a vegetable bed or serve as an aromatic edging for paths.

WHAT THEY NEED

These leafy crops thrive in a range of well-drained soils, but need consistent moisture to produce a lush head of foliage. Plants in containers and raised beds are vulnerable to drying out and should be watered regularly, before they show any sign of wilting, and grown in light shade during summer. Many salad leaves and annual herbs tend to bolt – rapidly grow upwards to flower – if they are dry at the roots. This causes leaves to become small and bitter. Some salad crops, such as rocket and oriental leaves, are also prone to bolting if sown too early in the year. Protect all of these tender leaves from slugs and snails, which can wipe out a row of seedlings overnight.

LETTUCES

Quick and easy to grow, and bursting with freshness when just picked, lettuce is a must-have crop that can be sown in the soil or containers almost year-round. Choose from the densely packed leaves of headed varieties or the more open, easily plucked foliage of loose-leaf types.

DIFFICULTY Easy

WHEN TO SOW Mar to July (standard varieties); Aug to Sept (winter varieties)

IDEAL SOIL TYPE Fertile, moist but well-drained soil

SITE REQUIREMENTS Sunny or slightly shaded in summer

GERMINATION TIME 7–10 days

GROW FROM Seeds or plants

YIELD 6–12 heads per 2m (6ft) row, depending on variety

CALENDAR

	WINTER			SPRING			SUMMER			AUTUMN		
SOW												
HARVEST												

Standard varieties

Late winter varieties

Time between sowing and harvesting
8–12 weeks

SOWING

Always use fresh seed, because old lettuce seed is unlikely to germinate well. Make small successional sowings every 2–3 weeks for continuous crops. Sow directly into the soil from early spring to midsummer by creating a shallow trench about 1cm (½in) deep using a cane. Sprinkle seeds thinly along the bottom of the trench, label the row, and cover seeds with soil. Water thoroughly using a watering can fitted with a rose. Space rows 15cm (6in) apart for smaller varieties and 30cm (12in) apart for larger headed lettuces. Seeds may not germinate well in hot weather, so make summer sowings in a lightly shaded spot or during the cool of the evening.

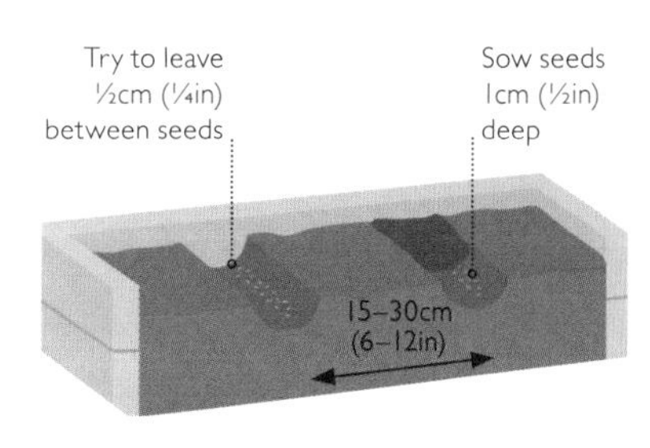

Direct sowing works well for lettuce seed as long as the soil has been raked into a fine tilth and the temperature is below about 25°C (77°F).

PLUG PLANTS If you only want a few lettuces or have difficulty protecting seedlings from slugs, raise plants in small pots or buy young plants and plant out from March until June. Sow two or three seeds per pot at the same depth as you would in the soil. Sow winter lettuce varieties in small pots from late summer to autumn so that they can be transplanted into a greenhouse.

GROW

WATER Water the soil around the plants regularly during dry weather and as they mature. Avoid splashing the leaves; this can cause problems with scorch or fungal diseases (see *pp.80–81*).

PROTECT Create traps and barriers to prevent slugs and snails from destroying seedlings. Check the crops at night with a torch and remove any pests. Cover lettuces with cloches in early spring and autumn to protect them from cold and extend the growing season. Grow winter lettuces in an unheated greenhouse.

THIN Thin seedlings out to their final spacing along the row when they are about 5cm (2in) tall: spacings vary from 15 to 30cm (6 to 12in) depending on the variety. Remove unwanted plants with your fingers and add them to salads as delicious baby leaves.

Slug traps can be made by pouring beer into a jar set level with the soil.

HARVEST

There's more than one way to pick a lettuce. Both headed and loose-leaf varieties can be cut when mature by using a sharp knife to slice through their stem at soil level. Loose-leaf varieties can be harvested gradually by picking just a few outer leaves from each plant as required. This makes more efficient use of space, as plants will crop over several weeks. Leaves are best eaten freshly picked, but can be stored for a short time in the fridge in a sealed plastic bag.

Mature lettuce heads are best harvested early in the morning, using a sharp knife to cut the base of the plant.

VARIETIES

Lettuce is available in a huge range of tastes, textures, and colours. There are two main forms –compact, dense varieties and loose-leaf varieties. Grow a selection to find your favourites.

'CATALOGNA' A vigorous loose-leaf variety, which forms open heads of long, serrated, green leaves.

'COCARDE' An attractive loose-leaf variety with a bronze tinge to its oak-shaped leaves.

'FRECKLES' This crisp semi-cos lettuce has leaves speckled with burgundy splashes.

'LITTLE GEM' This compact cos variety is quick to mature and produces dense hearts of crisp, sweet leaves.

'MARVEL OF FOUR SEASONS' A hardy butterhead lettuce, which can be sown successionally to crop year-round. Its pretty pale green leaves are dusted red.

'SALAD BOWL' This fast-cropping variety is ideal for beginners.

'TOM THUMB' This lettuce forms compact heads up to 13cm (5in) across, making it ideal for containers and small gardens.

WHY NOT TRY?

Any loose-leaf lettuce variety can be grown to produce cut-and-come-again baby salad leaves, which will be ready to harvest just 3–4 weeks after sowing and make efficient use of limited space. Sow seeds directly into a wide trench in the soil, about 1cm (½in) deep, or into a deep seed tray filled with multi-purpose compost. Aim to scatter seeds about 1cm (½in) apart. When plants reach about 5cm (2in) tall, cut the leaves with scissors as required, just above the central growing point at the base of the plant. Keep watering and leaves will regrow for a second – and sometimes even a third – harvest.

Lettuce leaves do not store well, so cut-and-come-again varieties are a great bet for a regular supply of fresh leaves.

RADICCHIO AND CURLY ENDIVE

Slightly bitter, crunchy leaves betray that radicchio and curly endive are chicories. Radicchio forms small, tight, burgundy heads, while curly endive has flatter, ruffled clusters of green leaves. They are grown in similar ways.

DIFFICULTY Easy
WHEN TO SOW Apr to July
IDEAL SOIL TYPE Moist, fertile, well-drained
SITE REQUIREMENTS Full sun or light shade
GERMINATION TIME 7–21 days
GROW FROM Seeds or plants
YIELD 6–8 heads from a 2m (6ft) row

CALENDAR

	WINTER	SPRING	SUMMER	AUTUMN
SOW				
HARVEST				

Time between sowing and harvesting
12–14 weeks

SOWING

Chicories can be sown from mid-spring for a summer harvest of leaves, but you might prefer to make summer sowings after an earlier crop, such as broad beans, has been lifted. Such sowings produce a welcome salad crop hardy enough to stand in the soil through autumn.

IN THE SOIL Use a rake to flatten the soil surface and break down any lumps. Fix a string line across the bed and draw out a shallow trench about 1cm (½in) deep across the bed. Sow, cover over with soil, label the row, and water well. Leave 30cm (12in) between the rows.

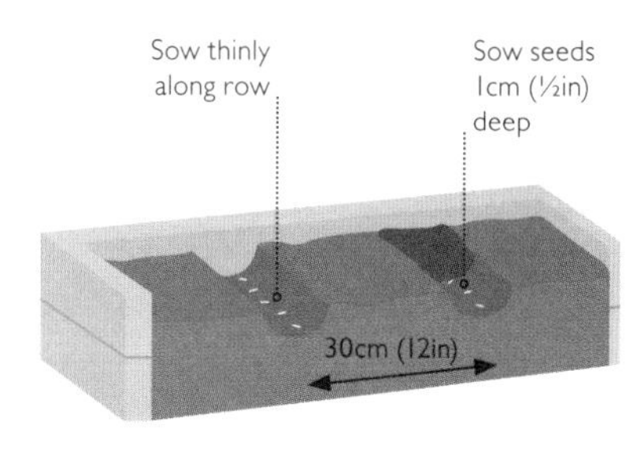

Sow seeds thinly along the trench using your thumb and forefinger..

IN POTS If sowing space is limited or you have a bad slug problem in your garden, you can raise seedlings in module trays or small pots. Fill the pots with multi-purpose compost, make a dent about 1cm (½in) deep at the centre of each and sow two seeds into it. Cover over with compost, label the pots, and water thoroughly. Place outdoors, on a windowsill or in a greenhouse to germinate. Young plants are also sometimes available to buy in summer.

GROW

THIN Gradually thin seedlings to 25–35cm (10–14in) apart, depending on variety. Uproot the small plants or cut the base of their stems with scissors and use the thinnings in salads. Harden off plants grown indoors in pots before planting out, by gradually acclimatizing them to outdoor conditions. Plant 25–35cm (10–14in) apart and water thoroughly.

WATER Plants in the soil require little attention, but need to be watered in hot, dry weather to prevent them flowering prematurely (bolting). Plants in containers benefit from regular watering. Avoid splashing the leaves to help prevent rotting. Weed around plants regularly with a hoe or hand fork.

Thin radicchio seedlings when they reach a manageable size.

PROTECT Slugs and snails may damage chicory crops, so place barriers and traps around plants to protect them. In cold regions, shield plants from frost by covering them with cloches or fleece tunnels to keep leaves in good condition during autumn and early winter. Provide as much ventilation as possible to help prevent rot setting in.

Frames covered with garden fleece provide protection from the cold.

HARVEST

Once heads have formed, harvest plants by cutting the stem just above ground level with a sharp knife or scissors. Leave stumps in place to regrow for a second harvest. Chicory is usually shredded and eaten raw in salads, but is also delicious baked or braised.

A Treviso-type radicchio is harvested with a clean scissor cut at its base.

Winter radicchio protects healthy hearts beneath frost-nipped outer leaves.

WHY NOT TRY?

Stopping light from reaching an endive makes its leaves paler and taste less bitter. This is called blanching. You can blanch an endive when the plant is mature, and only when its leaves are dry (to avoid problems with rotting). Lay a dinner plate, or other large, circular object, over the centre of a plant and leave it for 10–14 days until the leaves beneath are almost white. Regularly remove slugs that have taken up residence. Blanched heads need to be cut promptly because they don't keep for long.

Blanched endive has a slightly bitter taste, perfect with pear and blue cheese.

VARIETIES

There is not an huge range of varieties to choose from. Check hardiness and suitability for early or late sowings.

'CESARE' A modern radicchio variety, suitable for spring sowings, which forms dense, deep red hearts.

'DESPA' This endive has such a dense heart of frizzy leaves that it tends to self-blanch (*see left*) at the centre.

'PANCALIERI' This endive has superbly curled leaves with finely divided edges and suits all sowing times.

'ROSSA DI TREVISO' A striking narrow-leaved, upright radicchio that is exceptionally hardy.

PEPPERY LEAVES

Peppery cut-and-come-again leaves add a kick to any salad and are easy to grow. Ready to harvest in as little as four weeks, one plant will provide several cuts of baby leaves. While rocket is the most popular of these leaves, others are available and are grown in a similar way. If you like the zing of rocket, try mizuna, mibuna, and oriental mustard for a novel range of spicy flavours.

DIFFICULTY Easy
WHEN TO SOW Mar to Aug (outdoors); Sept (indoors)
IDEAL SOIL TYPE Well-drained and moist
SITE REQUIREMENTS Full sun, or light shade in summer
GERMINATION TIME 5–10 days
GROW FROM Seeds
YIELD 1kg (2lb) from a 2m (6ft) row

CALENDAR

	WINTER			SPRING			SUMMER			AUTUMN		
SOW				Outdoors	Outdoors	Outdoors	Outdoors	Outdoors	Outdoors	Indoors		
HARVEST	Indoors			Outdoors	Outdoors	Outdoors	Outdoors	Outdoors	Outdoors	Outdoors	Outdoors / Indoors	Outdoors / Indoors

Outdoors
Indoors

Time between sowing and harvesting
4–6 weeks

SOWING

Crops like rocket are sown densely and quickly yield a crop from a small space, making them ideal for filling gaps in beds and for growing in containers. When sown in hot weather these plants tend to bolt, or run to seed, prematurely, so are a good choice for intercropping, where they appreciate the light shade cast by taller vegetable plants.

IN THE SOIL Ensure the soil is thoroughly weeded and raked to remove any large lumps or stones. Anytime from early spring to late autumn, use a hoe to create a shallow, flat-bottomed trench about 10cm (4in) wide and 1cm (½in) deep. Scatter seeds in the trench so that they fall about 1cm (½in) apart, to avoid the need to thin seedlings. Cover with soil, label, and water well. Seeds can also be sown in narrow trenches at the same depth and distance apart. Sow successionally every 2–3 weeks, leaving 10cm (4in) between rows.

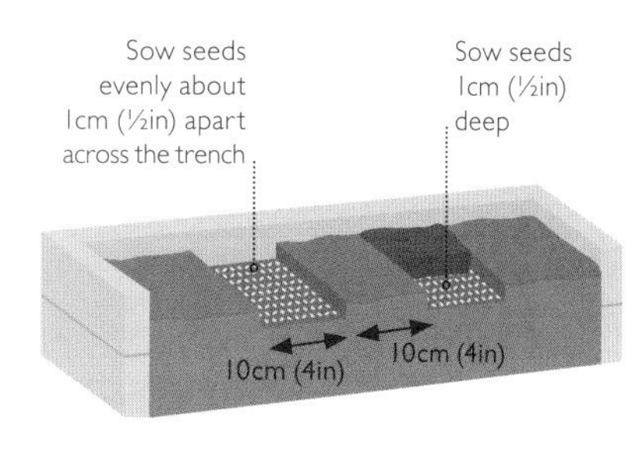

Rocket flowers are edible but are best removed to prolong leaf growth.

IN CONTAINERS Choose a wide container, such as a window box or trough, with a good surface area for sowing. Fill the container to about 3cm (1½in) from the top with multi-purpose compost, sow seeds 1cm (½in) apart across the whole surface, and cover with 1cm (½in) of compost. Label, water thoroughly, and place the container in full sun or light shade. Sowings made in early autumn and kept indoors on a windowsill should crop until early winter.

Sow rocket seeds thinly in a container. There is no need to thin the seedlings.

GROW

WATER Keep the soil moist to maintain growth and prevent bolting. Water plants growing in the ground regularly during summer. Plants in pots will need to be watered daily in hot weather and benefit from a feed with a balanced liquid fertilizer after their first cut.

PROTECT All of these plants are prone to attack by flea beetles, which eat little circular holes in the leaves. A small amount of damage is tolerable, but if the problem is severe, protect the plants by covering them with a fine insect mesh.

Covering your crop with fine mesh (with 0.8mm holes) excludes most pests.

HARVEST

Leaves are ready to harvest when they reach 8–10cm (3–4in) tall. The younger the leaves, the milder their flavour. Use scissors or a sharp knife to cut about 2cm (1in) above soil level leaving the central growing point intact, so the plants can regrow for further pickings.

To harvest small quantities, you can simply pick the outer leaves with your fingers as required. When plants begin to bolt or are growing in dry conditions they become more peppery until they taste quite bitter. The flowers are also edible. Mix your peppery favourites with lettuce and other mild leaves for wonderful varied salads.

Harvest peppery leaves in the morning while they are at their most succulent and eat as soon as possible.

WHY NOT TRY?

Try experimenting with a range of cut-and-come-again salads: there's a wide choice of colours, textures, and flavours to experience. Many leaf crops, including mild spinach and loose leaf lettuce, Swiss chard, bitter endive and chicory, crisp pak choi and komatsuna, and peppery mizuna and mibuna can be grown in exactly the same way as rocket, in the soil or containers. Sow them individually, buy one of the salad mixes available, or create your own combination by mixing seeds to sow together.

Once you've tried your own just-cut, homegrown baby leaf salad, you will never be tempted to buy a bag from the supermarket again.

Mibuna leaves have a peppery taste that adds a extra dimension to stir fries.

VARIETIES

Choose a single variety that appeals to your taste or sow a mixture full of diverse colours and flavours.

ROCKET 'ASTRA' Vigorous plants that resist bolting well and bear tasty, slender, serrated leaves.

ROCKET 'APOLLO' A salad rocket with rounded, soft leaves that develop rapidly and have a bold, peppery taste.

MIBUNA Produces reliable crops of smooth, upright leaves with a pale central stem and gentle peppery flavour.

MUSTARD 'RED GIANT' Attractive, spoon-shaped dark red leaves with a spicy, horseradish flavour.

SORREL

The citrus flavour of sorrel adds a distinctive zesty tang to salads, and its leaves also make a fabulous soup. Sorrel is easy to grow as a perennial plant, which dies down in winter and regrows each spring, but can be grown from seed each year if you would rather move it around the garden.

DIFFICULTY Easy
WHEN TO SOW Mar (indoors); Apr to June (outdoors)
IDEAL SOIL TYPE Well-drained and moist
SITE REQUIREMENTS Full sun or light shade
GERMINATION TIME 7–14 days
GROW FROM Seeds or young plants
YIELD Around six heads from a 2m (6ft) row

CALENDAR

	WINTER	SPRING	SUMMER	AUTUMN
SOW				
HARVEST				

From seed
Later years

Time between sowing and harvesting
8–10 weeks

SOWING

Sorrel's strong flavour means that few gardeners wish to grow a entire row of these plants. Sorrel seed can be sown thinly straight into the soil from mid-spring, at a depth of 1cm (½in), but for an early, more reliable crop, it's better to sow into small pots or modules from early spring to early summer and plant out seedlings later.

Fill the modules with multi-purpose compost, make 1cm (½in)-deep indents in the surface and place two seeds into each. Cover over with compost, water well, and label. Place on a sunny windowsill in cool spring weather; in mid-spring, place the seeds in a sunny spot outdoors to germinate.

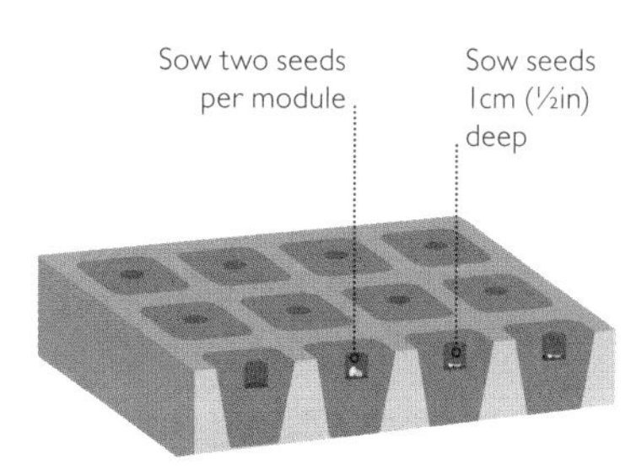

GROW

Plant out from late spring into a sunny or lightly shaded position, leaving 30cm (12in) between plants. Sorrel also grows well planted out into large containers. Water thoroughly after planting and continue to water regularly. Pinch off flower stems as soon as they appear during summer to encourage plants to keep producing leaves. The leaves will die back in late autumn, but sorrel is hardy enough to overwinter outdoors and regrow the following spring.

Common sorrel plants can reach 45cm (18in) in both height and spread.

HARVEST

Plants should have formed a dense cluster of foliage by mid-summer, when leaves can be pinched off with your fingers as required. Where sorrel is grown as a perennial and allowed to overwinter, new leaves will appear for picking from early spring.

VARIETIES

Consider flavour and appearance when making your selection of varieties.

FRENCH SORREL Also known as buckler-leaved sorrel, its small, shield-shaped leaves are perfect for salads.

COMMON SORREL A familiar broad-leaved sorrel, with upright, green, lemon-flavoured leaves that are best for cooking.

RED-VEINED SORREL A highly ornamental variety, with striking red veins dividing green lance-shaped leaves.

Red-veined sorrel makes a decorative feature in borders or large pots.

LEAF CELERY

A close relative of wild celery, this extremely hardy, robust plant is grown for its parsley-like leaves, which are used to add their sweet celery flavour to salads, stews, and soups. Leaf celery is a breeze to grow and a single sowing will provide generous pickings for many months.

DIFFICULTY Easy
WHEN TO SOW Mar to June (indoors)
IDEAL SOIL TYPE Well-drained and moist
SITE REQUIREMENTS Full sun or part shade
GERMINATION TIME 14–21 days
GROW FROM Seeds
YIELD Around eight heads from a 2m (6ft) row

CALENDAR

	WINTER			SPRING			SUMMER			AUTUMN		
SOW												
HARVEST												

Time between sowing and harvesting
10 weeks

Varieties like Par-cel are usually grown as annuals because they run quickly to seed.

SOWING

In early spring, sow seed indoors in small pots or modules. Place on a windowsill at 10–15°C (50–59°F); the plants will bolt at lower temperatures. Sow the seeds and cover thinly with compost or sand; they need light to germinate. Label, water well, and be patient, as germination is slow. A spring sowing will crop for a long period; sow again in mid-summer for a good winter harvest.

Sow the seeds thinly in small pots in spring, and ensure they don't get too cold.

GROW

Harden off young plants by acclimatizing them to outdoors conditions before planting out in full sun or light shade. Set them 25cm (10in) apart in rows 25cm (10in) apart, or find room for their lush, deep green leaves among herbs or in ornamental borders. Water well and protect from slugs and snails. Once established, leaf celery needs little care.

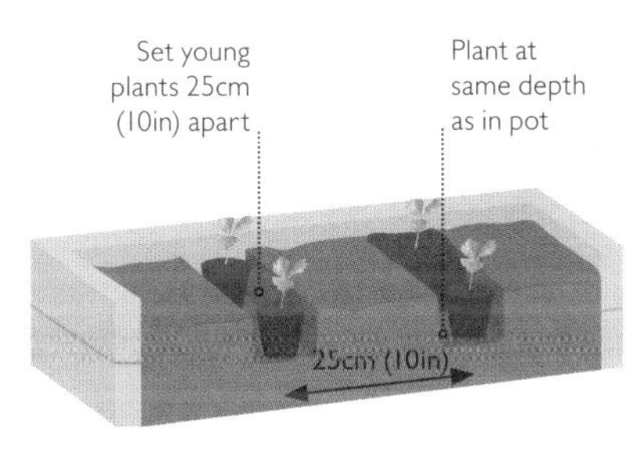

HARVEST

The first leaves should be ready to pick 4–6 weeks after planting out. Pick off the outer stalks with your fingers as required and new growth will arise from the centre. Leaf celery's hardiness means that it will stand outside for picking through winter, although there will be little new growth. Plants will run to seed and send up delicate flat heads of white flowers in spring.

VARIETIES

Your choice of variety will probably be based mainly on whether you prefer the texture of flat or curly leaves in your cooking.

LEAF CELERY The species plant (*Apium graveolens*) has flat, divided leaves and sturdy, slender stems.

'PAR-CEL' Widely available, the curly, dark green leaves of this variety are especially attractive.

PEA AND BEAN SHOOTS

Quick, nutritious, and bursting with fresh flavour and vitamin C, the young shoots of peas and broad beans can be harvested year-round on your windowsill. They are a great introduction to growing veg and will quickly become your go-to crops for pepping up salads and stir-fries.

DIFFICULTY Easy
WHEN TO SOW All year round (indoors)
IDEAL SOIL TYPE Well-drained and moist
SITE REQUIREMENTS Sunny windowsill
GERMINATION TIME 3–5 days
GROW FROM Seeds
YIELD 250g (9oz) from a standard seed tray

CALENDAR

	WINTER	SPRING	SUMMER	AUTUMN
SOW				
HARVEST				

Time between sowing and harvesting
2–3 weeks

SOWING

Soak seeds in cold water for 1–2 hours before planting; this will help to speed up germination. Choose a shallow container with a large surface area, such as a seed tray, and check that it fits on your windowsill. Fill your container with multi-purpose compost to about 2cm (1in) from the top. Drain the soaked seeds and scatter them thickly over the compost, with roughly 0.5cm (¼in) between each seed. Cover with 1cm (½in) of compost, water well, and allow to drain. Place the container on a warm, sunny windowsill with a tray underneath to catch any drips. Make a new sowing every four weeks for a plentiful supply of shoots all year round.

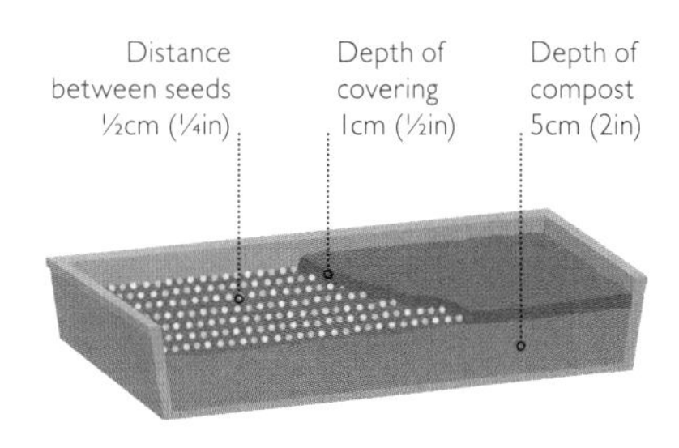

After scattering the seeds into a tray, cover them with a layer of compost about the same thickness as a pea.

GROW

Check the compost daily and water to prevent it drying out, but never leave the container standing in water. Once the dense mass of shoots begins to poke through the soil, turn the tray every day to stop the shoots from bending in one direction.

HARVEST

Harvest shoots when they are about 8cm (3in) tall. Pick a few for a garnish using your fingers or cut larger quantities with scissors. Stems cut above their first leaf will shoot away again for a second crop. When this second cut is finished, tip the container into the compost heap.

Use scissors to cut pea shoots just above their first leaf for a second harvest in less than two weeks.

VARIETIES

Any pea or broad bean seeds can be sown for shoots, but large quantities are needed. To keep costs down, use larger packs sold specifically for pea shoots, supermarket packs of dried peas, or dried broad beans (often sold as fava beans) from whole food stores.

CRESS AND MICROGREENS

Cress has long been cut as a seedling for garnishes, but a variety of leafy vegetables and herbs can be grown on a windowsill in the same way. Producing such "microgreens" is a great way to get a quick harvest from leftover seeds.

DIFFICULTY Easy
WHEN TO SOW All year round (indoors)
IDEAL SOIL TYPE Well-drained and moist
SITE REQUIREMENTS Sunny windowsill
GERMINATION TIME 2–5 days
GROW FROM Seeds
YIELD 150g (6oz) from a standard seed tray

CALENDAR

	WINTER			SPRING			SUMMER			AUTUMN		
SOW												
HARVEST												

Time between sowing and harvesting
1–3 weeks

SOWING

Grow microgreens in a seed tray or old plastic tub with drainage holes. Fill it with multi-purpose compost to about 1cm (½in) from the rim. Create a uniform, level surface to help achieve consistent germination, by removing any lumps from the compost and pressing the compost down gently using your hand. Water well. Sow seeds thickly and evenly across the tray, cover with about 0.5cm (¼in) of compost, and water again. Sow successionally every 2–4 weeks for a constant supply throughout the year.

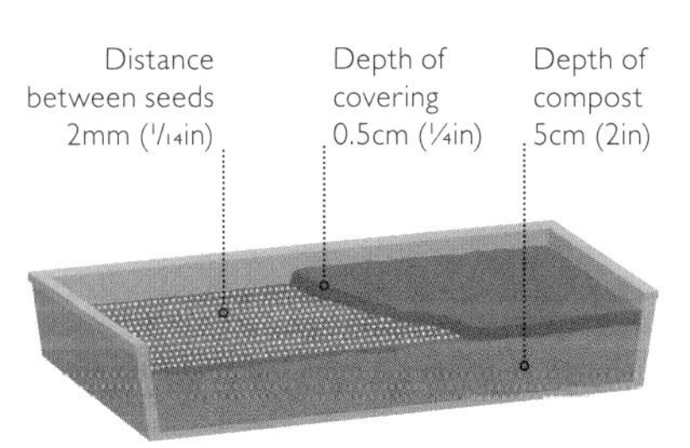

Water microgreens regularly using a watering can fitted with a rose.

GROW

Place the container on a warm, sunny windowsill with a tray underneath it to catch drips. Check the compost daily and water when the surface is dry. Never leave the container standing in water. Once the seedlings are growing, turn the tray daily.

HARVEST

Traditional salad cress is harvested when it is about 5cm (2in) tall and has just two seed leaves. Other microgreens are left to form one or two of their "true" leaves (which are like those on a mature plant) before picking. Cut the stems at the base with scissors as required and use immediately in salads and sandwiches, and as colourful garnishes.

VARIETIES

Have fun growing a range of leafy vegetables and herbs as microgreens to see what you enjoy. Go for crops with distinctive flavours and colours.

AMARANTH 'RED AZTEC' Gorgeous rich red seedlings with a gentle earthy taste.
BASIL A strong, sweet aniseed flavour in green or red varieties.
CORIANDER Fresh green shoots with an intense, slightly citrusy flavour.
KALE 'RED RUSSIAN' Violet-veined leaves have a subtle kale flavour.
MUSTARD 'RED GIANT' A punchy taste of horseradish and pretty dark red colour.
SALAD ROCKET 'UBER' Green seedlings with a powerful peppery kick.
SWISS CHARD 'BRIGHT LIGHTS' Multi-coloured stems that provide a subtle earthy flavour.

LEAFY HERBS

When bought from a shop, herbs can be pricey, so it pays to grow these kitchen essentials by your back door. Parsley, coriander, chervil, dill, and basil are annuals or biennials, with soft stems and aromatic foliage. They can be sown repeatedly through the year for a steady supply.

DIFFICULTY Easy
WHEN TO SOW Mar (indoors); Apr to late summer (outdoors)
IDEAL SOIL TYPE Well-drained and moist
SITE REQUIREMENTS Full sun or part shade
GERMINATION TIME 7–21 days
GROW FROM Seeds or plants
YIELD Around 500g (18oz) from a 2m (6ft) row

CALENDAR

	WINTER			SPRING			SUMMER			AUTUMN		
SOW				Indoors	Outdoors	Outdoors	Outdoors	Outdoors	Outdoors			
HARVEST	Outdoors / Indoors				Indoors	Outdoors / Indoors	Outdoors / Indoors	Outdoors / Indoors	Outdoors / Indoors	Outdoors / Indoors	Outdoors / Indoors	Outdoors / Indoors

Sown indoors
Sown outdoors

Time between sowing and harvesting
4–10 weeks

SOWING

Young plants are available to buy, but if you use herbs in quantity then it's easy to grow them from seed. Make successional sowings every 2–4 weeks from spring until early autumn, for a seamless supply. Most of these herbs are hardy, but basil is killed by frost and rots in cold, wet weather, so only plant it outdoors during summer and grow in pots on the windowsill in cool areas.

SOW IN MODULES Make early spring sowings into modules or small pots on a warm, sunny windowsill for planting out in mid-spring. Fill modules with multi-purpose compost, sow three seeds per module, cover lightly with a 5mm (¼in) layer of compost, label, and water. Keep the compost moist and remove the weaker seedlings where more than one germinates in a module. Continue to sow in this way to supply young plants to plant out throughout spring and summer.

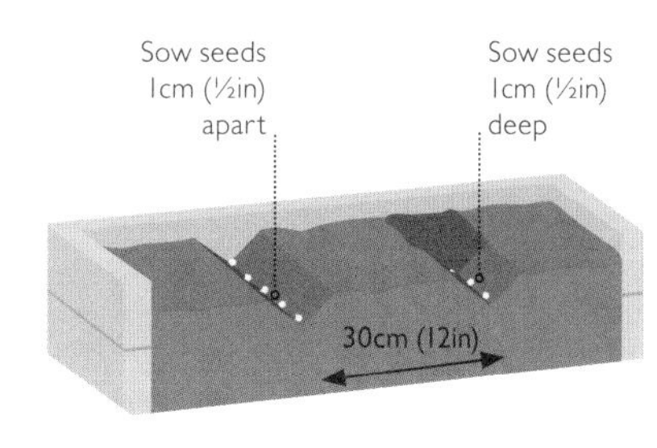

Sow herb seeds in half-size or narrow module trays that have been designed to fit perfectly on a windowsill.

SOW INTO THE SOIL From mid-spring, once the soil is warm, sow hardy herbs outdoors into raked soil. Wait until early summer to sow basil. Make spring sowings in full sun, but find a little shade for summer sowings of coriander and chervil; this will help prevent them bolting. Stretch a string line across the bed and use a hoe to make a shallow trench 1cm (½in) deep. Sow seeds thinly, about 1cm (½in) apart, along the trench, cover over with soil, label, and water well. Leave 30cm (12in) between rows. Herbs can also be sown at the same depth in large pots, window boxes, or old fruit crates.

Sow your favourite herbs as edging to a bed or between rows of vegetables.

GROW

PLANT OUT OR THIN Harden off young plants raised indoors by acclimatizing them to outdoor conditions. Plant into the soil in rows 30cm (12in) apart, at their final distances (*see box, right*) or gradually thin out rows of seedlings to the same spacings.

PLANTING DISTANCES
- **PARSLEY** 15cm (6in) apart
- **BASIL** 20cm (8in) apart
- **CORIANDER** 5cm (2in) apart
- **CHERVIL** 23cm (9in) apart
- **DILL** 20cm (8in) apart

WATER Keep soil moist to encourage leafy growth and help prevent plants running to seed. Water plants in containers daily in summer and check those growing in the soil regularly. To help prevent rot developing, water in the morning so that plants aren't left in cold, wet soil overnight. Herbs in containers will benefit from a balanced liquid feed applied once a week.

Harden off young parsley by leaving plants outside for progressively longer periods over a span of two weeks.

Coriander tends to run to seed rapidly, but regular watering encourages it to produce lush crops of leaves for longer.

HARVEST

Once your herb plants are growing strongly, regular picking will help to stimulate new, bushier growth and keep plants producing leaves rather than flowers. Leaves will always be at their plumpest in the morning, after the dew has dried, but are best picked fresh, just before use.

Pick outer stems from parsley and chervil; cut coriander to 5cm (2in) above soil level and let it regrow; pinch off the tips of basil shoots just above a leaf joint; and snip dill leaves as required. Pinch out flower stems to keep leaf production going for longer. You can capture some of the flavour of these herbs for use in the winter by freezing leaves in ice cube trays or making flavoured oils and vinegars. Note that the soft foliage of leafy herbs does not dry well.

Mixed green and purple basil makes for a decorative and fragrant display.

VARIETIES

Most herbs are available in numerous varieties, each with a specific flavour and growing requirements. It is best to start with the most familiar forms and explore from there.

BASIL 'NAPOLITANO' Known as lettuce leaf basil thanks to its large, crinkled, green foliage.

BASIL 'RUBIN' Glorious dark purple leaves, but particularly prone to rot in wet summers.

CHERVIL The fern-like leaves of the species *Anthriscus cerefolium* need no improvement by plant breeders.

DILL 'HERA' Produces dense foliage and is slow to bolt.

CORIANDER 'LEISURE' A large-leaved variety with good resistance to bolting.

PARSLEY 'MOSS CURLED' Tightly curled leaves have a slightly spiky appearance and are robustly savoury.

PARSLEY 'ITALIAN GIANT' A hardy, flat-leaved, strongly flavoured variety.

LEAFY PERENNIAL HERBS

The soft stems of perennial mint, tarragon, and fennel die down over winter, but regrow every spring to produce crops of fragrant foliage. These vigorous plants look good in herb gardens, ornamental beds, or containers.

DIFFICULTY Easy
WHEN TO SOW/PLANT Mar to June
IDEAL SOIL TYPE Well-drained
SITE REQUIREMENTS Full sun or light shade
GERMINATION TIME 7–14 days
GROW FROM Seeds (fennel), plants (mint, tarragon)
YIELD About 500g (18oz) from a 2m (6ft) row

CALENDAR

	WINTER	SPRING	SUMMER	AUTUMN
SOW/PLANT				
HARVEST				

Time between sowing and harvesting
6–8 weeks

SOWING

Mint and tarragon are best grown from plants. Fennel should be grown from seed sown directly into the soil outdoors because its seedlings do not transplant well. Sow fennel seed in a sunny location as soon as the soil warms up in spring; this herb requires well-drained soil. Sow by making a trench 1cm (½in) deep across a bed using a hoe or trowel. Place 3–4 seeds at 30cm (12in) intervals, cover with soil, label, and water well. Thin to one seedling at each point where several emerge.

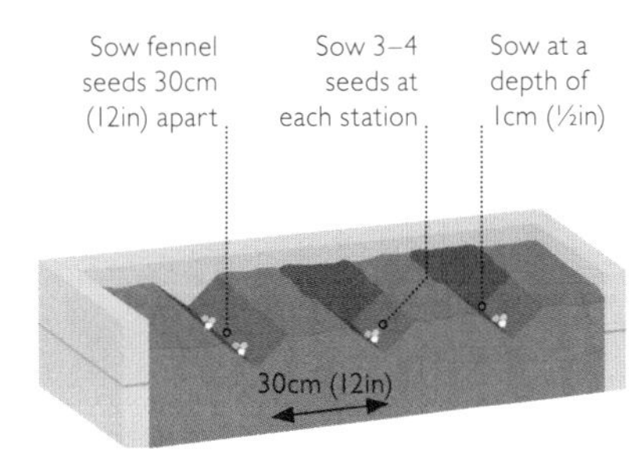

GROW

PLANT OUT Buy young mint and French tarragon plants from a nursery or garden centre and plant them in spring or early summer. Allow a space of 30cm (12in) between mint plants and 60cm (2ft) between tarragon plants. The roots of mint spread rapidly, to the point where the plant can be invasive if not kept in check; for this reason, grow mint in pots, or restrict its spread. To do this, take a large plastic container, cut off its base, and sink it into the soil so that just its rim protrudes. Plant the mint within this container.

Plant tarragon into fertile, free-draining soil or in a large pot filled with gritty compost. It makes an attractive foliage plant, ideal in a herb border or group of pots. It is tall when mature, so is best at the back of a bed or group of pots.

Mint spreads rapidly by underground runners. Contain it in a sunken plastic pot to stop it from becoming a nuisance.

Tarragon grows well in containers. Place plants in a sunny position where they are sheltered from wind.

WATER Herbs growing in containers need regular watering in spring and summer, along with feeding with a balanced liquid fertilizer to promote lush foliage. Mint prefers moist soil and should be watered daily if in a container. Once established, tarragon and fennel growing in open soil will need watering only during extended periods of hot, dry weather.

Fennel needs watering during extended dry periods.

CUT BACK As summer arrives, mint and fennel will flower and their leaves will coarsen. To encourage a new flush of foliage that will be ready to harvest well into autumn, cut the plants back in June; mint should be cut close to the base and fennel to 30cm (12in). Feed and water the plants. You can leave fennel uncut to enjoy its pretty umbels of yellow flowers, but make sure to remove them before seeds are scattered all over the garden.

French tarragon tends not to flower and only needs to be trimmed occasionally to keep it tidy.

Feathery foliage and pretty flowers make fennel a decorative garden addition.

HARVEST

Cut fresh herbs as required. Mint is perfect with all kinds of early summer veg and made into tea; the delicate aniseed taste of fennel works well with fish; while French tarragon has a stronger sweet aniseed flavour that's famously good paired with chicken. Cut the dried heads of fennel to harvest the delicious seeds. These herbs are best enjoyed fresh, but their tastes can be preserved in vinegars and butter.

VARIETIES

A bewildering range of mint varieties exists, while little choice is on offer when it comes to tarragon and fennel.

'BOWLES'S MINT' Tall, vigorous stems with rounded, softly furry leaves.

'CHOCOLATE MINT' Attractive dark brown-tinged leaves and stems.

'SPEARMINT' Smooth, slightly toothed leaves that are versatile in the kitchen.

FRENCH TARRAGON Long stems of upright, slender, anise-flavoured leaves. Avoid inferior Russian tarragon.

WHY NOT TRY?

Aromatic Mediterranean herbs, including thyme, sage, and rosemary, are easy-to-grow perennials with woody stems. They thrive in sunny spots with free-draining soil, where they will not need watering once established, and also make forgiving pot plants that can survive irregular watering. They're usually bought as young plants in spring, although thyme grows well from seed sown during spring into pots of sandy compost. Harvest by picking the soft new growth at the tips of stems as required. Trim plants back after flowering to keep them compact.

Mediterranean herbs are perfect additions to an ornamental garden. Just one or two plants will provide ample pickings for the kitchen.

TROUBLESHOOTING

Beds bursting with fresh salad leaves look fantastic, but all that succulent foliage attracts pests that can wipe out whole rows of crops if left unchecked. Tender leaves are also particularly vulnerable to fungal diseases. Learn to recognize problems and act quickly to prevent damage. Providing the correct growing conditions makes your plants more resilient.

APHIDS

PROBLEM Brown patches and stunted growth; visible clusters of aphids.
CAUSE Aphids puncturing leaves to feed on the plant's sap.
REMEDY Monitor crops and squash any aphids between your fingers to prevent them multiplying. Where damage is minimal, aphids can easily be washed off leaves before eating.

LETTUCE ROOT APHIDS

PROBLEM Established lettuces suddenly wilt and die during summer.
CAUSE Sap-sucking insects feeding on the roots.
REMEDY Dig up and destroy affected lettuces; keep healthy plants well watered. Grow lettuces under fine insect mesh during summer or choose varieties bred with resistance.

CARROT FLY

PROBLEM Young parsley plants discolour, wilt, and grow poorly or die.
CAUSE The small, white larvae of carrot fly feeding on thick parsley roots.
REMEDY Avoid growing parsley in soil where carrots or parsnips have recently been grown. Keep plants well watered so they establish quickly; mature plants can cope with some root damage.

FLEA BEETLE

PROBLEM Tiny circular holes in leaves of rocket, mizuna, and other members of the cabbage family.
CAUSE Small beetles feeding.
REMEDY Sow seeds into warm soil and water well to promote strong, resilient growth. Cover crops with fine insect mesh. Tidy up dead leaves in autumn to prevent beetles overwintering.

SLUGS AND SNAILS

PROBLEM Ragged holes in leaves; seedlings chewed down to soil level.
CAUSE Feeding of slugs and snails, especially at night or in wet weather.
REMEDY Protect seedlings using sawdust or eggshell barriers, set beer traps, and put copper tape around containers. Check under containers and patrol around crops with a torch at night.

BOLTING

PROBLEM Plants rapidly grow upwards to flower, rather than producing leaves.
CAUSE Changes in day length or stressful growing conditions, such as dry soil.
REMEDY Water to keep soil moist throughout growth. Be careful to sow at the right time of year, because some crops are sensitive to day length and will bolt if sown too early or late.

DAMPING OFF

PROBLEM Seedlings sown indoors collapse, sometimes with fuzzy fungal growth visible.
CAUSE Soil-borne fungi.
REMEDY Sow into clean pots and fresh compost. Sow seed thinly and provide good ventilation to keep air circulating and reduce humidity, because the fungi thrive in moist conditions.

LETTUCE DOWNY MILDEW

PROBLEM Yellow patches on lettuce leaves; fuzzy growth on underside of leaves, which eventually turn brown.
CAUSE Fungal infection, which is worse in wet conditions.
REMEDY Pick off affected leaves and do not compost them. Grow plants further apart to increase airflow; avoid splashing foliage when watering.

MINT RUST

PROBLEM Mint plants produce distorted growth coated in an orange, powdery layer. Affected leaves may die and fall.
CAUSE A fungal disease that affects all varieties of mint.
REMEDY Remove infected plants, including their spreading roots, as soon as possible, and plant mint in a new location.

POWDERY MILDEW

PROBLEM A white, dusty coating on the leaves of herbs and occasionally salads.
CAUSE A fungal infection, more common in dry conditions.
REMEDY Water regularly to keep the soil moist and plants growing vigorously. Improve air circulation around plants by growing them further apart. Remove infected leaves that fall onto the soil.

Borlotti beans are shelling beans, which are meant to grow and dry on the pod before being picked. However they are also delicious when eaten young.

PEAS AND BEANS

Nutritious and packed with protein, homegrown peas and beans reliably deliver sweet, fresh flavours. Luxuriant foliage, attractive flowers, and hanging pods make them decorative plants and their large seeds are fun for children to sow.

THE TASTE OF SUMMER

Snapping open the first pods of tender peas and broad beans and gathering handfuls of slender French beans are highlights of early summer. Once you've enjoyed the first harvest, keeping up with these prolific plants becomes the challenge; it's no great hardship, though, because young, sweet pods are so convenient for snacking on outdoors and easy to use in the kitchen. Juicy runner beans are just as productive, but mature slightly later. Extend the season into autumn by protecting late sowings of dwarf French beans under fleece and planting exquisitely pink-flecked borlotti beans and other gourmet varieties to dry and store for feasting on during winter.

VERTICAL HARVEST

Peas and beans are large, vigorous plants, which need to be grown in groups of at least five or six for a worthwhile harvest. Nevertheless, they are surprisingly easy to fit into almost any space. The trick is to clothe every bit of vertical space in your plot – a bare garage wall or ugly fence – with fast-growing, free-flowering climbing varieties, or build a support from canes to screen bins, create privacy, or use as a tall feature. Where this isn't an option, many dwarf varieties of peas and beans are available; these attractive bushy plants still produce impressive yields, and do well in windier locations. Both climbing and dwarf varieties will flourish in large, well-watered containers.

IDEAL CONDITIONS

As members of the legume family of plants, peas and beans have the extraordinary ability to fix nitrogen from the air and use it to fuel growth, thanks to root nodules filled with friendly bacteria. The nodule-laden roots are usually left in the soil when plants are removed to add fertility for the next crop. That's no excuse for not improving the soil with plenty of compost, however, because this helps retain moisture around roots in hot summer weather. Be aware when sowing, that peas and broad beans can withstand frost, while French and runner beans cannot, and need warmth to germinate.

PEAS, MANGETOUT, AND SUGAR SNAPS

The pop of a just-picked pea pod and the sugary taste of its contents are pleasures exclusively reserved for those who grow their own, because they quickly fade after harvest. Given fertile soil and shelter from strong winds, it's simple to grow attractive rows of these climbing plants up pea sticks for a bumper crop.

DIFFICULTY Easy

WHEN TO SOW Oct to Nov, and Feb to Mar (early varieties); Mar to July (maincrop varieties)

IDEAL SOIL TYPE Fertile, moist but well-drained

SITE REQUIREMENTS Sunny and sheltered

GERMINATION TIME 7–28 days

GROW FROM Seeds or plants

YIELD About 2kg (4lb) from a 2m (6ft) row

CALENDAR

	WINTER			SPRING			SUMMER			AUTUMN		
SOW			Early	Early	Maincrop	Maincrop	Maincrop	Maincrop			Early	Early
HARVEST					Early	Early	Early	Maincrop	Maincrop	Maincrop	Maincrop	Maincrop

Early varieties
Maincrop varieties

Time between sowing and harvesting
12–16 weeks

SOWING

Peas fall into two main categories: early and maincrop varieties. Maincrop peas take longer to grow but produce heavier yields. Mangetout and sugar snaps are varieties eaten young in the pod.

Sow seeds outdoors by placing them in a trench or pushing them into the soil with your finger to a depth of 2.5–4cm (1–1½in). Plant two parallel rows 20cm (8in) apart to allow room for supports in between. Space seeds 5cm (2in) apart along the rows for dwarf varieties and 10cm (4in) apart for tall varieties.

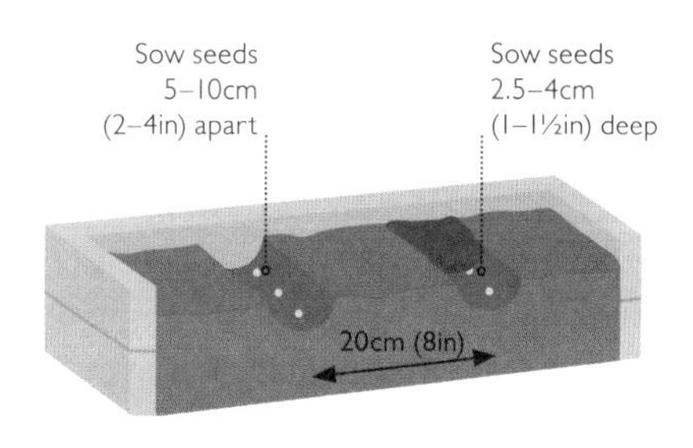

EARLY VARIETIES In milder areas, faster-maturing early varieties are hardy enough to tough out the winter when sown in late autumn, but in colder regions you should sow them in early spring, once the soil begins to warm. Germination can be unreliable from early sowings, so warm the soil under cloches or horticultural fleece for a week before sowing, or sow your first batch in modules or a length of guttering in an unheated greenhouse to plant out in mid-spring.

MAINCROP VARIETIES Maincrop peas can also be sown from early spring. They take a few more weeks to mature than early varieties, so if sown at the same time should continue your harvest later into summer. If you have space, make a sowing each month until midsummer. Opt for plants over seeds if you have heavy soil or if there's a risk of mice eating the germinating seeds.

Sowing in guttering in a greenhouse helps prevent damage by mice.

Warm soil under horticultural fleece before planting early pea varieties.

GROW

STAKE Before or just after sowing, push sturdy, branched sticks (known as pea sticks) firmly into the soil along the rows, or secure a length of netting or chicken wire between posts for the tendrils of young plants to curl around.

WATER During their early growth, peas only need watering in dry periods. If they are thirsty while flowering, however, it will reduce your crop, so soak the soil at the base of flowering plants weekly if there is no rain.

PROTECT Mice often feast on germinating pea seeds, which may make it necessary to raise seedlings in pots. Cover rows of seedlings with netting in early spring to stop pigeons stripping the young shoots.

Branched pea sticks are a natural support, easily grabbed by pea tendrils.

HARVEST

Pick the pods of shelling peas when they are fat, but still bright green. Older pods are rougher and lose this vibrant colour, and the peas inside become hard and starchy rather than sweet. Check carefully among foliage for concealed pods. Mangetout pods are ready when the outline of the peas inside is just visible. Sugar snap pods are best picked when smooth and rounded. Eat all peas as soon as possible after picking, because they quickly lose their sweetness.

Netting peas deters pigeons, but unravelling nets to weed is a chore.

Harvest peas in the morning, when their sugar content is at its highest.

WHY NOT TRY?

Pea shoots are a tasty addition to salads and are incredibly quick to grow on a windowsill year-round. Fill a wide seed tray with multi-purpose compost to about 2cm (1in) from the top and cover the surface with seeds so that they are roughly 5mm (¼in) apart. Sowing this densely takes a lot of seed, so use leftover seeds from last year or larger, cheaper packets sold specifically for pea shoots. Cover with 1cm (½in) of compost and water the tray thoroughly using a watering can fitted with a rose; allow the compost to drain and place on a sunny windowsill. Water regularly to keep the compost moist. The shoots should be ready to harvest in about 3 weeks, when they are about 6cm (2½in) tall.

Cut pea shoots just above the lowest leaf to let a new crop regrow.

VARIETIES

Dwarf pea varieties can be grown without supports. Snap and mangetout peas are eaten in the pods, but can also be allowed to grow longer and shelled as normal.

'KELVEDON WONDER' A popular dwarf early variety for autumn or spring sowing.

'METEOR' Extremely hardy, dwarf, early pea that will crop well in colder areas.

'ONWARD' A maincrop pea that yields stocky pods packed with large peas.

'OREGON SUGAR POD' A popular tall maincrop mangetout variety.

'RONDO' This maincrop variety produces heavy crops of long pods.

FRENCH BEANS

A taste of summer, just-picked French beans have a fresh, green flavour and a satisfying snap. Both climbing and bushy dwarf varieties are attractive plants, particularly when dripping with green, purple, or yellow pods. If picked regularly, they will crop heavily for up to two months.

DIFFICULTY Easy

WHEN TO SOW Apr (under cover); May–July (outdoors)

IDEAL SOIL TYPE Fertile, moist, and well-drained

SITE REQUIREMENTS Sheltered, full sun

GERMINATION TIME 7–14 days

GROW FROM Seeds or plants

YIELD About 2kg (4½lb) from a 2m (6ft) row

CALENDAR

	WINTER			SPRING			SUMMER			AUTUMN		
SOW												
HARVEST												

Under cover
Outdoors

Time between sowing and harvesting
60–70 days

SOWING

Dwarf and climbing varieties are sown in the same way. Climbing varieties need fertile, moisture-retentive soil to produce a bumper crop. Provide this by digging a trench to a spade's depth in late winter, adding a layer of well-rotted organic matter, and covering over with soil so it's ready to sow later in spring.

WARM THE SOIL Never sow French beans into cold soil. They need a minimum soil temperature of 10°C (53°F) to germinate. Warm the soil under cloches or horticultural fleece for two weeks before spring sowings. Buy young plants if you live in a colder area where germination outdoors is difficult.

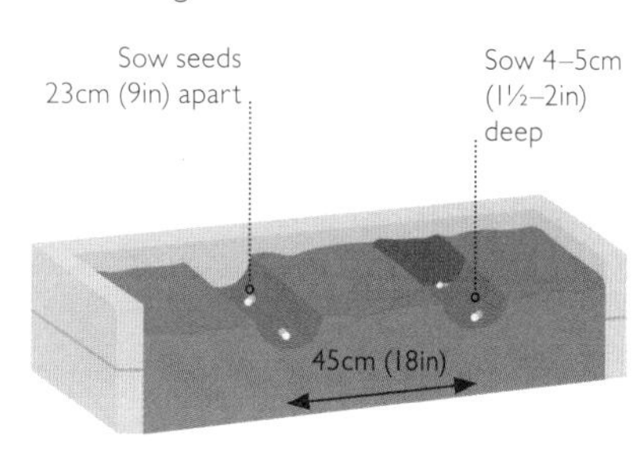

SPACING Sow dwarf beans 23cm (9in) apart along rows with 45cm (18in) between them, or in blocks with plants 30cm (12in) apart in every direction. Start climbing varieties at the base of a circle of supporting canes, which can be put in place before or after sowing. Sow two seeds 4–5cm (1½–2in) deep at each point, to help ensure successful germination.

French beans also grow well in containers; grow a single dwarf plant in a 20cm (8in) diameter pot, or sow seeds 23cm (9in) apart in larger pots. Two sowings 3–4 weeks apart should supply all the beans you need.

If the weather is cold, start seeds under cover in pots on a windowsill.

GROW

PROTECT Harden off young plants that you have bought or raised indoors before planting out. Cover early sowings with cloches or tunnels in cold spring weather, ensuring good ventilation, as still air encourages fungal diseases. Extend your harvest into autumn by covering dwarf varieties with cloches or tunnels as the weather cools.

SUPPORT Add a tripod of canes 2.5m (8ft) high soon after sowing climbing varieties. Tie young plants to supports and pinch out the growing tip of each plant when it reaches the top to direct energy to pod production. Prop up dwarf varieties with branched sticks to stop pods from touching the soil.

Lightly tie beans to supports, taking care not to bruise the stems.

French beans are traditionally grown supported by a double row of canes; tripod supports are better if space is limited.

WATER Water plants often, particularly during flowering, to encourage beans to set and the formation of further flowers. French beans are self-pollinating, so insect pollination is not necessary for pods to form.

Water well during summer; a compost mulch at the base helps retain water.

HARVEST

Pick young pods regularly, while they are tender and stringless, to encourage plants to keep cropping. Check carefully among the leaves, because sometimes pods are well hidden, and snap off the thin stalks with your fingers. Young pods are delicious eaten raw or lightly steamed as soon as possible after picking. French beans freeze well once blanched in boiling water for 30 seconds.

Keep picking beans throughout the seasons; if you let beans ripen on the plant, it will cease production of new pods.

VARIETIES

French beans are available in rounded, fine, or flat-podded varieties, in shades of purple, green, and yellow. Dwarf varieties crop quicker, but climbers bear pods for longer.

'BLAUHILDE' This vigorous climber produces very long, flattened, purple pods that turn green when cooked.

'COBRA' A trusted climbing variety that produces reliable early crops of long, rounded, green pods; its mauve flowers are highly decorative.

'PURPLE TEEPEE' A dwarf variety with dark purple pods that turn green when cooked.

'SAFARI' This cultivar produces reliable crops of tasty, stringless green beans on dwarf plants

'SONESTA' The waxy, butter-yellow pods produced look decorative on this compact dwarf variety.

RUNNER BEANS

These statuesque climbers are worth growing for their vibrant scarlet, bi-coloured, or white flowers, but also yield a plentiful harvest of long, juicy pods through summer and well into autumn. Pick the pods when young, while they are tender and their distinctive flavour is at its best.

DIFFICULTY Easy

WHEN TO SOW Apr (indoors); May to June (outdoors)

IDEAL SOIL TYPE Deep, fertile, well-drained, and moist

SITE REQUIREMENTS Sheltered in full sun or part shade

GERMINATION TIME 5–10 days

GROW FROM Seeds or plants

YIELD 2kg (4½lb) from a 2m (6ft) row

CALENDAR

	WINTER			SPRING			SUMMER			AUTUMN		
SOW					Indoors	Outdoors	Outdoors					
HARVEST								Outdoors / Indoors	Outdoors / Indoors	Outdoors / Indoors	Outdoors / Indoors	

Indoors
Outdoors

Time between sowing and harvesting
10 weeks

SOWING

In late spring or early summer, sow two beans 5cm (2in) deep at the base of each supporting pole (*see below*). To give plants a head-start, you can sow in mid-spring in deep pots indoors on a warm windowsill for planting out later.

PREPARE THE SOIL These plants need moist, fertile soil. Dig a trench to a spade's depth in late winter, covering its base with a 10cm (4in) layer of well rotted organic matter; back-fill with soil. Warmth is needed for germination, so don't sow too early into cold, wet soil.

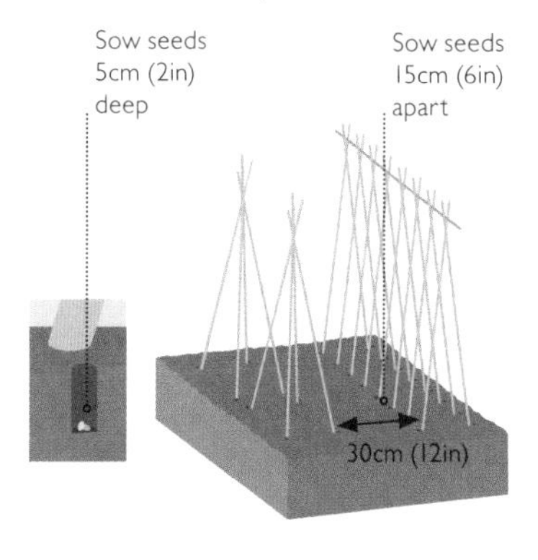

In cool areas, warm the soil by covering with cloches or horticultural fleece for two weeks before sowing. Runner beans can also be grown in containers that are deep enough to support poles.

Cane pyramids are suitable supports in smaller spaces, or when growing in pots.

SUPPORTS Build sturdy supports before sowing; they will need to carry the weight of these vigorous plants, which can reach 2.5m (8ft) tall or more. Try pyramids made from four to six 2.5–2.7m (8–9ft) bamboo canes pushed firmly into the soil and tied securely where they cross; alternatively use A-frames – two rows of crossing poles pushed into the ground at an angle and tied to a horizontal pole laid across the top. Leave 15cm (6in) between each pole and at least 30cm (12in) between rows. Beans can also be trained up arches, netting, and trellis.

A-frames are ideal supports for larger plantings of runner beans.

GROW

PROMOTE CLIMBING When the bean seedlings have germinated, thin them to one plant per pole. Harden off any seedlings you have raised indoors by gradually acclimatizing them to outdoor conditions and only plant them out once the risk of frost has passed.

Plant some ornamental flowers nearby to help attract the insects needed to pollinate runner bean flowers. Carefully tie the stems of growing plants to the poles using twine, to encourage them to climb. Pinch out the growing tip of each plant when it reaches the top of its pole.

WATER Runner beans are thirsty plants that need to be watered well during any spells of dry weather, but particularly when their flowers start to form because this will help beans to set. Evening watering seems to work best. Once they are climbing quickly, add a 5cm- (2in-) thick mulch of well-rotted compost around plants, to help retain moisture in the soil.

Runner beans are self-fertile but need to be pollinated by insects.

HARVEST

Runner bean pods can reach more than 30cm (12in) in length, but long ones tend to become stringy. They are best picked young at about half that length. Regular picking stimulates plants to produce further flowers and can keep them cropping until the first frosts in autumn. Eat runner beans raw or lightly cooked as soon as possible after harvest. Freeze excess beans after blanching them in boiling water for 30 seconds.

Pick beans every 3–4 days; the more you pick, the more will grow.

WHY NOT TRY?

Dwarf varieties are a perfect alternative to climbing plants where space is at a premium or conditions are too windy. These bushy plants grow to a height and spread of 45–60cm (18–24in). It's better to grow them raised up in pots rather than in the soil, as beans trailing on the ground will be damaged by slugs. Water pots daily and apply liquid feed weekly during summer to keep plants healthy. Dwarf varieties crop for a shorter period than climbers, so make a second sowing in mid-summer to continue picking pods into autumn. 'Hestia' is a widely available variety with pretty red and white flowers.

Dwarf runner beans such as 'Hestia' put on a great floral display.

VARIETIES

'FIRESTORM' This modern, self-pollinating, red-flowered variety is known for reliably setting heavy crops of stringless pods.

'LADY DI' Red flowers produce long, fleshy pods that develop slowly and remain good to pick for a long period.

'MOONLIGHT' This self-pollinating variety is white-flowered and consistently produces excellent crops, even in hot, dry summers.

'SCARLET EMPEROR' This trusted old variety is widely grown for its scarlet flowers and bumper harvests.

'ST GEORGE' Attractive red and white bi-coloured flowers precede heavy clusters of long, tender beans.

'WHITE LADY' Popular for its dependably high yields of tasty beans following abundant white flowers.

BROAD BEANS

Among the first crops to appear through the soil in spring, these plants can reach a height of 1.2m (4ft) but are little or no trouble to grow. Their unusual black and white flowers appear along the main stem, followed by a succession of fat, green pods for several weeks in summer.

DIFFICULTY Easy
WHEN TO SOW Oct–Nov (hardy varieties); Feb–Apr (all varieties)
IDEAL SOIL TYPE Well-drained, fertile
SITE REQUIREMENTS Shelter, full sun or part shade.
GERMINATION TIME 7–14 days
GROW FROM Seeds or plants
YIELD ABOUT 2kg (4½lb) from a 2m (6ft) row

CALENDAR

	WINTER	SPRING	SUMMER	AUTUMN
SOW				
HARVEST				

Hardy varieties
All varieties

Time between sowing and harvesting
12–16 weeks

SOWING

Hardy varieties of broad beans are among the first crops that can be sown outdoors. As long as the soil is well-drained and not frozen, they can be sown in late winter, and will germinate despite the cold. In mild areas, hardy varieties can be sown in late autumn for a slightly earlier crop in late spring, although overwintering these young plants is a bit of a gamble because they may not survive severe winters; in particularly cold spells, you will need to protect the plants under cloches or horticultural fleece. Further sowings can also be made in spring for harvests later in the summer.

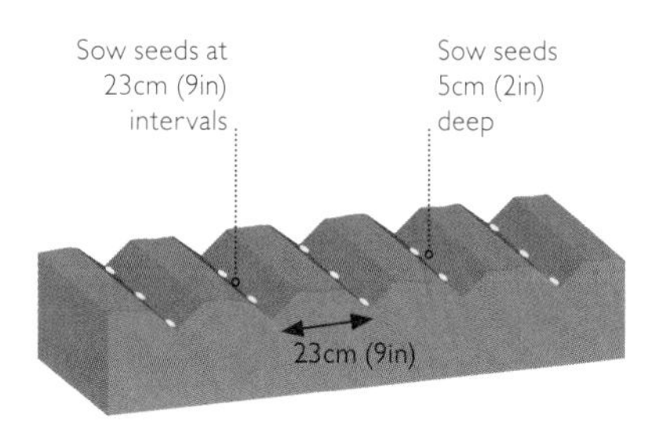

Before sowing outdoors fork over the soil and add compost or well-rotted manure.

IN THE SOIL Mark out two rows, 23cm (9in) apart, with a string line and use the corner of a hoe or a trowel to draw out a trench about 5cm (2in) deep along each. Sow the large seeds individually at 23cm (9in) intervals, staggering the two rows to make the most of the growing space. Sow a few extra seeds at the end of each row to produce spare plants to fill any gaps. Cover over the rows with soil and water well.

IN POTS Raise young plants in pots where winters are cold (hungry mice steal seeds) or for a head start in early spring. Choose pots or modules tall enough to sow seeds 5cm (2in) deep, fill them with multi-purpose compost, and push one seed into each pot. Water well and place on a sunny windowsill, in an unheated greenhouse, or in a sheltered spot outdoors to germinate. Harden off and plant out in spring, at the same spacing as for sowing seeds. Choose more compact dwarf varieties to sow directly into large containers.

Large bean seeds need deep modules for sowing under cover.

GROW

WATER Broad beans in containers or raised beds need regular watering, but those in the ground only need it during hot, dry weather. Spread a 5cm- (2in-) thick mulch of well-rotted compost around plants in spring to help retain soil moisture.

PROTECT Add supports by tying string between strong canes at the end of each row. This will keep taller varieties upright in windy summers. Dwarf varieties don't need supports. When plants are flowering, but before pods form, pinch out about 7.5cm (3in) from the top of the stem, with your fingers or scissors, to help pods swell.

Cutting off the tops of the stems stops blackfly from infesting young shoots.

A web of garden twine woven around a frame of canes helps protect bean plants from weather damage.

HARVEST

The leafy tops pinched out in spring make delicious steamed greens. Tender young beans become starchier as they grow, so it's best to pick young pods regularly, just as they turn to point downwards on the plant and the beans are about size of your thumb nail. Pick pods from the base of the stem upwards, by gently twisting them off the plant. When the crop is finished, cut plants down at the base to leave their nitrogen-rich roots to break down in the soil.

Excess beans can be successfully frozen immediately after picking.

VARIETIES

Choose the hardiest varieties for autumn and late winter sowings and dwarf varieties for containers and exposed plots.

'AQUADULCE CLAUDIA' Hardy and productive, ideal for sowing in autumn and late winter.

'CRIMSON FLOWERED' Beautiful rich-red flowers precede good crops of pods.

'RED EPICURE' Unusual deep red beans that can be eaten raw when tiny.

'THE SUTTON' A sturdy dwarf variety for autumn or spring sowings.

'WITKIEM MANITA' Reliably fast-maturing from spring sowings.

DRYING BEANS

Beans like borlotti, with their beautiful pink-speckled pods, are known as drying beans. The pods are left on the plant until they become hard and the beans rattle inside, when they are shelled and stored for later use. While borlotti are the most popular drying beans, other varieties are available and are grown in a similar way.

WHEN TO SOW Apr (under cover); May (outdoors)

IDEAL SOIL TYPE Fertile, moist, and well-drained

SITE REQUIREMENTS Sheltered, full sun

GERMINATION TIME 7–14 days

GROW FROM Seeds or plants

YIELD About 2kg (4½lb) from a 2m (6ft) row

CALENDAR

	WINTER			SPRING			SUMMER			AUTUMN		
SOW					Under cover	Outdoors	Outdoors					
HARVEST							Under cover	Under cover		Outdoors	Outdoors	

Under cover
Outdoors

Time between sowing and harvesting
65–90 days

SOWING

Climbing varieties are vigorous and need fertile, moisture-retentive soil. Dig a trench to a spade's depth in late winter, add well-rotted organic matter, and cover with soil, ready for sowing in spring. Thereafter, dwarf and climbing varieties are sown in the same way.

AN EARLY START It is desirable to sow borlotti early in cooler areas because the beans mature slowly. However, note that they are tender plants and will not tolerate frost. Never sow them into cold soil but wait instead for mild weather and warm the soil under cloches or horticultural fleece for two weeks before sowing. Alternatively, start seeds in pots on a windowsill indoors. They should be ready to plant out about 7–8 weeks later when they are about 8cm (3in) tall.

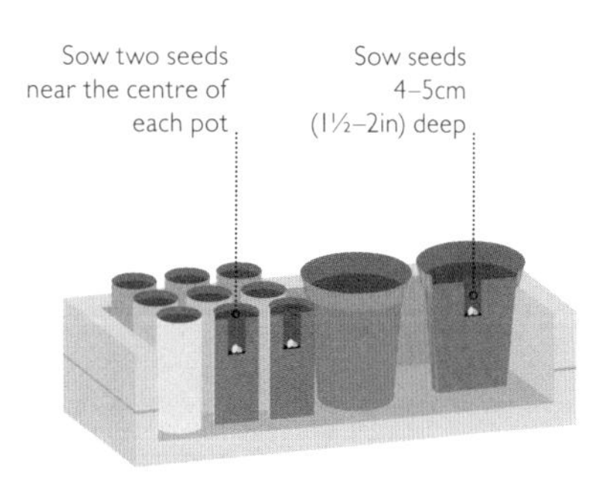

Borlotti are best sown in pots under cover. This will also protect seedlings from slugs and snails, which love young leaves.

SPACING If sowing outdoors, sow pairs of seeds 4–5cm (1½–2in) deep to ensure that one plant germinates at each point. Space dwarf varieties at 23cm (9in) intervals along rows 45cm (18in) apart, or in blocks with plants 30cm (12in) apart. Start climbing varieties about 23cm (9in) apart at the base of a circle of supporting canes, which can be put in place before or after sowing. If starting under cover, sow two seeds into each deep pot.

Sow seed outdoors only after the last frost has passed and the soil is warm.

GROW

PROTECT Harden off any young plants that you have raised indoors gradually before planting them out. In cold spring weather, cover young plants with cloches or tunnels at night. Protect them with netting to keep birds off and set mouse traps if beans are being eaten.

SUPPORT Add a pyramid or some other sturdy arrangement of 2.5m- (8ft-) tall canes after you have sown the seed of climbing varieties. Tie the young plants in to the supports to encourage them to climb, and pinch out the growing tip of each plant when it reaches the top of the cane. Use branched sticks to support dwarf varieties and so prevent pods touching the soil.

WATER Water plants regularly, particularly during flowering, to encourage beans to set and the formation of further flowers. Stop watering in late summer to allow the plants to die back and the pods to dry.

Plant out seedlings into rich soil when they are big enough to handle.

Use spare woody branches as supports for dwarf bean varieties.

HARVEST

Borlotti beans can be popped from fresh pods to use in cooking but are more usually dried. To harvest dried beans, leave pods on the plant until they become hard and the beans rattle inside. Pick in warm, dry weather, then shell and store in air-tight containers. In wet conditions, bring pods indoors and lay on newspaper for about a week to dry. Only store beans that are fully dry and always cook them before eating.

Split open the inside seam of the pod to remove the beans.

WHY NOT TRY?

A popular Japanese snack, edamame beans are the immature pods of the soya bean, and in regions with hot summers it's easy to grow your own. Sow seeds in pots or trays on a windowsill in spring. Once germinated, continue to grow on the windowsill, moving plants into larger pots as necessary, until they can be hardened off and planted out in warm summer weather. Space plants 15cm (6in) apart, in rows 45cm (18in) apart, in full sun and well-drained soil, and by late summer their branches will be laden with green edamame pods.

Steam edamame pods then enjoy the tender protein-packed beans.

VARIETIES

Grow a range of drying beans to provide a variety of colourful ingredients to add to hearty soups and stews in winter.

'BORLOTTO LINGUA DI FUOCO' Available in climbing and dwarf varieties, both with gorgeous pink-speckled pods and beans.

'CANADIAN WONDER' This dwarf bean produces plentiful pods that can be eaten whole when young or left to dry.

'LAMON' Plentiful pods filled with gourmet beans are produced on climbing plants.

'YIN YANG' (ORCA BEAN) Ideal for containers, this dwarf plant yields attractive black and white dried beans.

'SPAGNA BIANCO' A climbing bean with long flat pods to eat green or dry.

TROUBLESHOOTING

Given good conditions, peas and beans shoot up so fast that they outgrow most problems. They are most vulnerable after sowing outdoors and as seedlings, because the large nutritious seeds are a magnet for pests. Grow peas and beans in a different area of the garden each year to help prevent pests building up in the soil.

BEAN SEED FLY

PROBLEM Seedlings of French and runner beans emerge damaged and distorted or are killed.
CAUSE Maggots of the bean seed fly eating the roots of seedlings.
REMEDY Sow seed into warm soil – rapid development allows seedlings to outgrow damage; or sow beans in pots to plant outdoors once growing strongly.

BLACK BEAN APHID

PROBLEM Clusters of small black insects gathering to feed on young shoots, weakening plant growth.
CAUSE Sap-sucking aphids.
REMEDY Pinch out the succulent young tips of broad bean plants before they are attacked. Be vigilant and squash any aphids with your fingers as soon as you see them.

MICE

PROBLEM Seeds disappear from the soil or from ripening pods, especially those left on the plant to dry.
CAUSE Rodents feeding.
REMEDY Sow seeds in pots kept on a windowsill indoors, in a greenhouse, or off the ground outdoors, for planting out later. Set traps if the problem is particularly severe.

PEA AND BEAN WEEVIL

PROBLEM Distinctive, small, U-shaped notches cut out of the edges of pea and bean foliage.
CAUSE Tiny insects feeding on leaves.
REMEDY Tolerate damage where plants continue to grow well. Where damage is serious, try excluding weevils and boosting growth by growing crops under horticultural fleece.

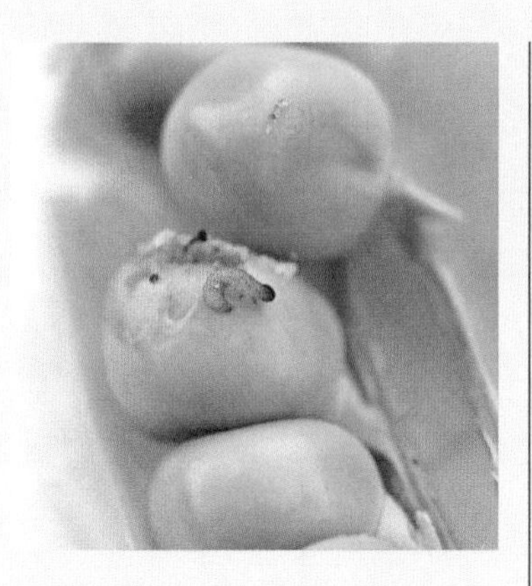

PEA MOTH

PROBLEM Tiny caterpillars seen on peas when the pods are opened.
CAUSE Feeding b y moth larvae that emerge from eggs laid on pea flowers.
REMEDY Sow peas in early spring or early summer to avoid having peas in flower while the moths are flying. If pea moth is especially troublesome, grow crops under fine insect mesh.

PIGEONS

PROBLEM Young foliage stripped; drying seeds eaten.
CAUSE Wood pigeons feeding on young shoots (particularly of peas) or pecking at pods to eat drying seeds.
REMEDY Cover the crop with netting or fleece. Branched twigs stuck into the soil and angled over seedlings also provide some protection.

SLUGS AND SNAILS

PROBLEM Seedlings or pods damaged or destroyed, usually overnight.
CAUSE Slugs and snails feeding.
REMEDY Protect seedlings using sawdust or eggshell barriers, set beer traps, and encircle containers with copper tape. Support plants to prevent pods touching the soil. Go out after dark and manually remove the pests.

CHOCOLATE SPOT

PROBLEM Circular brown spots on the leaves of broad beans.
CAUSE A fungal disease most common in wet or humid weather.
REMEDY Improve air flow by removing any crop coverings. Next year, leave more space between plants to improve air circulation. Always water the soil rather than the foliage.

NO BEANS SETTING

PROBLEM Flowers drop from plants leaving no tiny pods developing.
CAUSE A lack of pollinating insects; hot weather, or insufficient moisture.
REMEDY Add plenty of organic matter before planting, mulch around plants to retain moisture and water regularly. Try white flowered runner bean varieties reputed to set better crops in heat.

POWDERY MILDEW

PROBLEM A white, talc-like covering on foliage; plant growth slows.
CAUSE A fungal disease that infects plants weakened by dry conditions.
REMEDY Water regularly to keep the soil moist; mildly affected plants may recover. Add more organic matter to the soil before future plantings and mulch crops to retain moisture.

Shallots are grown in the same way as onions, and store just as well through winter, but their smaller bulbs have a far more nuanced, aromatic flavour.

THE ONION FAMILY

Vegetables from the onion family, also known as alliums, are at the heart of many everyday dishes and are essential crops for any garden. Although mostly slow to mature, they are easy to grow and worth the wait for their rich flavours.

PATIENCE PAYS

For a late summer bounty of onions and shallots, be ready to sow seed or plant miniature bulbs (called sets) in spring. Once this is done, however, there is little more work involved: no successional sowing is needed, and there's no rush to pick onions while they are young – just lift them to dry in the last summer sunshine, before storing for autumn and winter. Garlic and leeks both need even longer in the ground to mature, but like onions, garlic stores well once dried, while leeks are hardy enough to stand in the soil through winter, where they look magnificent and are always ready to unearth for soups and stews. Make a few sowings of quick-cropping spring onions to keep your tastebuds tingling while you wait.

WHERE TO GROW?

Alliums come in all sizes, from compact chives and spring onions, which will thrive in a patio pot, to enormous walking-stick onions that need open soil to reach their full potential. Although onions and leeks will grow perfectly well in containers, a worthwhile harvest is only really possible where there is space to grow them in quantity in the soil. Clusters of shallots, garlic, and spring onions are a better bet for a satisfying crop from pots, but their slender foliage will need backup from other leafier crops to create an attractive display.

WHAT THEY NEED

These crops all need free-draining soil that has been improved with organic matter in autumn, and an open, sunny position. The fleshy roots of onions, shallots, and garlic are prone to rotting in wet conditions and excessive feeding and watering encourages soft growth that makes this more likely. Leeks, on the other hand, love a rich soil and will thrive given consistent watering and a thick mulch once they're growing strongly. Regular weeding is the final key to success, because the upright, narrow leaves of alliums do nothing to shade out weeds. Pay particular attention to keeping the tiny, spindly, slow-growing seedlings weed free, to get them off to the best possible start.

ONIONS

Onions are a valuable staple and can be grown from seed or baby bulbs called "sets". Starting from seed is cheap, but means a longer growing season and potentially more problems while the slender seedlings establish. Sets are pricier but more robust, making them the best choice for beginners.

DIFFICULTY Easy from sets, moderate from seed

WHEN TO SOW Feb (seed indoors); Mar to Apr (seed and sets outdoors)

IDEAL SOIL TYPE Fertile, well-drained with a pH of over 6.5 (*see p.20*)

SITE REQUIREMENTS Full sun

GERMINATION TIME 7–21 days

GROW FROM Seeds or sets

YIELD 4kg (8lb) from a 2m (6ft row)

CALENDAR

	WINTER			SPRING			SUMMER			AUTUMN		
SOW			Seed indoors	Seed/sets outdoors	Seed/sets outdoors							
HARVEST								■	■	■	■	

Seed indoors
Seed/sets outdoors

Time between sowing and harvesting
4–6 months

SOWING

Avoid sowing or planting in cold soil, because growth will be poor and the onion plants may flower too early or "bolt". Warm the soil under cloches or horticultural fleece in early spring before sowing, or try raising seedlings indoors on a cool windowsill. Sets can be planted as soon as the soil warms up and becomes workable in spring.

SETS Create a trench 2.5cm (1in) deep. Gently push sets into the trench, spaced 5–10cm (2–4in) apart, with their pointed tips upwards. Push the soil back around them with your hands so that just the tips are showing; firm the soil gently. Water if the soil is dry.

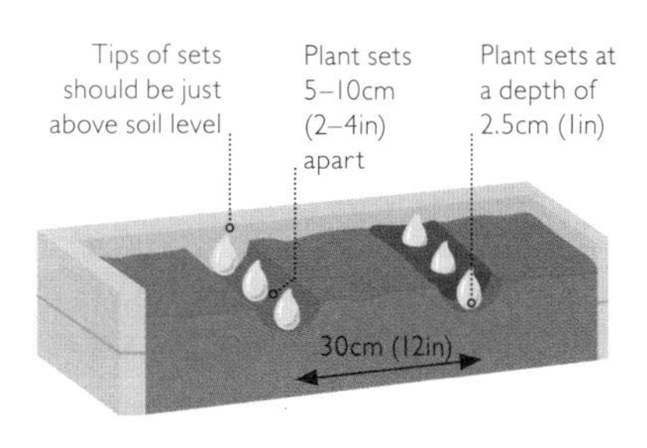

Onion sets are onions that have deliberately not been grown to full size.

SEED Sow seed indoors in late winter in a modular seed tray filled with compost to about 1cm (½in) from the top. Sow 3 5 seeds into each module, and cover with compost to fill. Water and place on a cool windowsill at 10–16°C (50–61°F). Acclimatize to outdoor conditions before planting clumps of 3–5 plants spaced 25cm (10in) apart in mid-spring. Alternatively, sow outdoors in early to mid-spring. Sow seeds thinly about 1cm (½in) deep in rows 30cm (12in) apart, cover with soil, and water using a watering can fitted with a rose.

GROW

WEED Tiny onion seedlings do not compete well with weeds, so weed rows regularly. Be careful not to damage developing bulbs with sharp tools.

WATER Water regularly in dry weather, but avoid overwatering, particularly as bulbs start to swell, as this can slow growth and encourage fungal diseases.

THIN Gradually remove weaker seedlings to leave plants spaced 8–13cm (3–5in) apart along the row. Use thinnings as spring onions in salads.

PROTECT Birds may pull sets out of the ground. These can be replanted, but if the problem persists, cover the row with netting. Where onion fly is a pest (*see p.92*), grow under fine insect mesh.

Remove weeds from young onion beds regularly by hand to prevent damage.

HARVEST

From midsummer, onions can be lifted gently with a fork to be used fresh. Store the bulk of your crop by lifting in late summer or autumn, when the leaves are more yellow than green. Lay them out on the ground to dry in the sun for about 10 days or bring them under cover in wet weather. When the outer skin is papery, store in net bags or plait dry leaves together.

Onions planted in clumps make best use of your growing space.

Onions can be eaten as soon as they have been lifted, or left to dry outdoors or in a cool, well-ventilated place indoors.

WHY NOT TRY?

Tree onions make a wonderful addition to a garden, with their 1m- (3ft-) tall stems topped with miniature onion bulbs instead of flowers. When stems eventually fold down under the weight, the little bulbs root where they land, allowing the plant to migrate each year, giving it the name "walking onion". Plant the marble-sized bulbs in spring or autumn, in the same way as onion sets, and they will flower in the summer of the second growing season. The small bulbs are ideal for pickling; the leaves can be used like chives, while the large bulbs at the base can be harvested as cooking onions.

Tree onions have an unusual form making them a talking point.

VARIETIES

Varieties with yellow skins have a robust flavour, red bulbs are sweeter, while white-skinned types have a delicate flavour. The choice of varieties is more limited for sets than it is for seeds.

'HYRED' Rounded, deep red bulbs full of sweetness, with good resistance to bolting.

'RED BARON' Slightly flattened, red-skinned onions have a strong taste and store well.

'SETTON' This reliable, high yielding, yellow-skinned variety stores well.

'SNOWBALL' The pale, thin-skinned bulbs of this white onion have a mild taste.

'STURON' This widely available yellow-skinned onion produces medium-sized bulbs that keep well.

SHALLOTS

Shallots look like small onions, but have a distinct slightly sweet flavour; they can be yellow- or red-skinned, rounded or elongated. Clusters of up to ten bulbs form from each set, allowing a single row to yield a generous summer harvest, which will store well until the following spring.

DIFFICULTY Easy
WHEN TO SOW Nov to Mar (sets); Mar (seed)
IDEAL SOIL TYPE Fertile, well-drained with a pH>6.5
SITE REQUIREMENTS Full sun
GERMINATION TIME 7–21 days
GROW FROM Seeds or sets
YIELD About 1.5kg (3lb 5oz) from a 2m (6ft) row

CALENDAR

	WINTER			SPRING			SUMMER			AUTUMN		
SOW	Sets	Sets	Sets	Sets, Seeds								
HARVEST								Sets, Seeds	Sets, Seeds			

Sets
Seeds

Time between sowing and harvesting
20–24 weeks

SOWING

Growing from sets is the easiest option, but starting from seed is cheaper. Sow seed indoors in early spring, 1cm (½in) deep in small modules. Place on a cool windowsill, and water well. Alternatively, sow outdoors as soon as the soil warms in spring. Rake to create a fine surface, draw out a row 1cm (½in) deep and sow along it thinly. Leave 10cm (4in) between rows.

Planting times for sets vary, so follow the supplier's instructions. Make a trench 2.5cm (1in) deep and place sets along it at 15cm (6in) intervals, with their pointed tips upward; firm the soil around them. Water thoroughly if the soil is dry. Space rows 23cm (9in) apart.

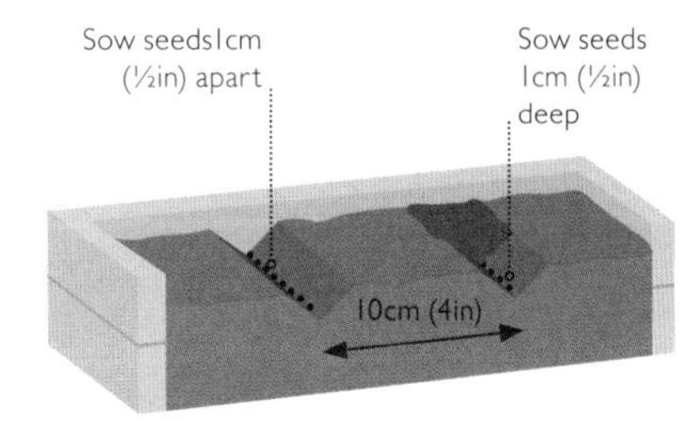

GROW

Harden off seedlings raised indoors and plant them 5cm (2in) apart in rows separated by 10cm (4in). Thin seedlings sown outdoors to the same spacing. Weed among rows regularly, especially while seedlings are small. Only water during dry weather. Protect rows of sets with netting if birds pull them up.

Shallots have shallow roots; weed around them by hand, not with a hoe.

HARVEST

Lift shallots gently with a fork when their foliage begins to yellow and bend downwards. Remove soil around the roots and leave them on the soil to dry. In wet weather, dry on racks or newspaper indoors. Once dry, hang them in bunches or net bags to store at room temperature.

Correctly dried shallots will keep for up to 12 months.

VARIETIES

There's a good choice of varieties whether you choose seeds or sets.

'AMBITION' Blush-skinned, rounded variety with a strong flavour. From seed.

'GOLDEN GOURMET' Produces heavy crops of rounded, yellow-skinned bulbs. From sets.

'ZEBRUNE' Long banana shallots with a mild taste. From seed.

SPRING ONIONS

By far the quickest allium crop, spring onions can be ready to liven up dishes just eight weeks after sowing. Given well-drained conditions, successional sowings will flourish in raised beds, patio containers, or on a windowsill, making picking possible for nine months of the year.

DIFFICULTY Easy
WHEN TO SOW Mar to July; Aug to Sept (overwintered)
IDEAL SOIL TYPE Well-drained
SITE REQUIREMENTS Full sun
GERMINATION TIME 7–14 days
GROW FROM Seeds
YIELD 60–70 onions from a 2m (6ft) row

CALENDAR

	WINTER			SPRING			SUMMER			AUTUMN		
SOW				Normal	Normal	Normal	Normal	Normal	Overwintered	Overwintered		
HARVEST	Normal				Overwintered	Overwintered	Normal	Normal	Normal	Normal	Normal	Normal

Normal sowing
Overwintered

Time between sowing and harvesting
8–12 weeks

SOWING

Begin sowing outdoors as soon as the soil warms in spring. Sow thinly in narrow drills 1cm (½in) deep and 15cm (6in) apart. Cover with soil, label, and water thoroughly. Make successional sowings every three weeks until a final sowing in late summer or early autumn to overwinter and pick in spring.

Spring onions also grow well in containers on a windowsill or in a sunny spot outdoors. Fill a large pot or window box with multi-purpose compost to about 3cm (1½in) below its rim, sow seeds thinly over the surface, cover with a 1cm (½in) layer of compost and water thoroughly.

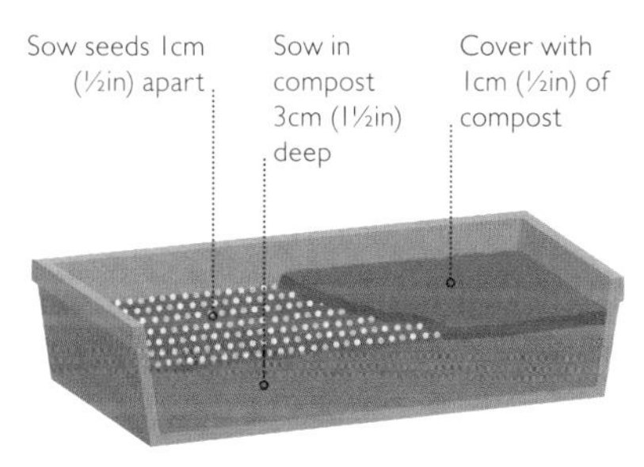

GROW

Slender spring onions can be crowded out, so prevent competition by weeding regularly at every stage of growth. Thin seedlings to 2–3cm (1–1½in) apart; add the thinnings to salads for a surprisingly strong burst of onion flavour. Water plants in containers regularly to prevent compost drying out; crops in the soil only need watering during dry weather. Protect overwintering crops with cloches to ensure a spring harvest.

Thin spring onions seedlings when they are about 5cm (2in) high.

HARVEST

Pull up spring onions as soon as they reach a useable size, loosening the soil with a fork if their roots have a firm grip. Harvest any onions that bolt as soon as possible and eat their flower stems too. Spring onions are best eaten freshly picked; their green foliage is just as delicious as the white stem.

Spring onions may have white or more ornamental red stems.

VARIETIES

Both white and red varieties crop well through summer and autumn.

APACHE A crisp, red variety for spring and summer sowings.

'GUARDSMAN' This slender, straight white variety suits autumn and spring sowings.

'WHITE LISBON' A popular, fast-growing white old variety with a strong flavour.

GARLIC

Although it evokes the scents of the Mediterranean, garlic is actually very hardy and best planted in autumn to overwinter outdoors for harvesting in summer. Plant plenty, because the fat, juicy bulbs are full of aromatic appeal when freshly unearthed, and can easily be stored well into winter.

DIFFICULTY Easy
WHEN TO SOW Oct to Nov
IDEAL SOIL TYPE Well-drained
SITE REQUIREMENTS Full sun
GERMINATION TIME 4–6 weeks
GROW FROM Cloves
YIELD 18–20 bulbs from a 2m (6ft) row

CALENDAR

	WINTER	SPRING	SUMMER	AUTUMN
SOW				
HARVEST				

Green garlic
Mature garlic

Time between sowing and harvesting
5–9 months

PLANTING

Buy garlic as bulbs and split them into cloves for planting. You should ideally plant in autumn because garlic needs about six weeks of cold temperatures if it is to divide into bulbs: otherwise you'll get one large clove-less bulb at harvest.

In well-drained soil, use a trowel or dibber to make holes 10cm (4in) deep and drop a clove into each, with its pointed end upwards, before covering over with soil. Space plants 18cm (7in) apart in a block, or at 10cm (4in) intervals. Where conditions are wet, plant cloves in small pots and overwinter outdoors to plant out in early spring.

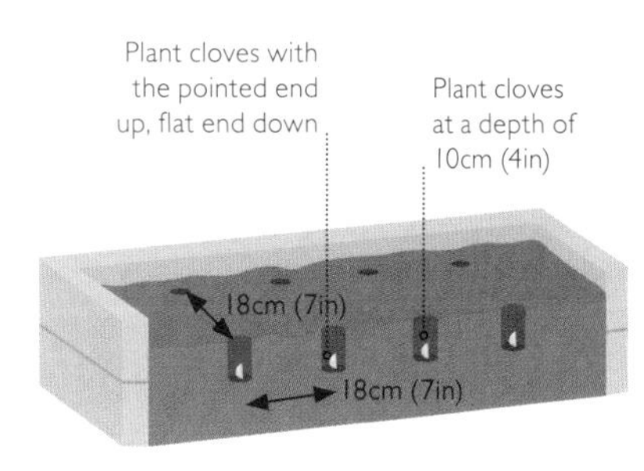

Hand weeding around young garlic shoots is often the only way to avoid damaging the tender leaves.

GROW

Weed carefully among the young plants with a hoe or by hand. Garlic bulbs swell during spring and will benefit from weekly watering during dry weather at this time. Keep garlic in containers well watered and apply a balanced liquid fertilizer weekly during spring. Cut back any flowering stems in late spring to help direct energy to the bulbs.

HARVEST

Unearth a few bulbs as "green" garlic in early summer to enjoy their unique mild flavour. Harvest the main crop when the leaves begin to yellow, by loosening the soil with a fork and pulling the stem. Lay plants on the soil to dry for a week, then hang bulbs by their stems.

Store dried bulbs in a dry, frost-free place after harvest. They will keep for six months or more.

VARIETIES

"Hard neck" varieties have the finest flavour, while "soft neck" varieties are the best choice for storing.

'LAUTREC WIGHT' This French hard neck variety is renowned for its rich flavour.

'PICARDY WIGHT' Ideal for storing, this soft neck variety has a fierce flavour for committed garlic enthusiasts.

CHIVES

Chives are a trouble-free, perennial herb with a powerful oniony flavour. Their clumps of upright, bright leaves, topped with purple pompom flowers in late spring, are equally at home in an ornamental garden, containers, or a vegetable patch; this makes chives ideal to plant as edging for a path.

DIFFICULTY Easy
WHEN TO SOW March to May
IDEAL SOIL TYPE Well-drained and moist
SITE REQUIREMENTS Full sun or light shade
GERMINATION TIME 7–14 days
GROW FROM Seeds or plants
YIELD 500g (1lb) from a 2m (6ft) row

CALENDAR

	WINTER	SPRING	SUMMER	AUTUMN
SOW				
HARVEST				

First year
Established plants

Time between sowing and harvesting
8–10 weeks

SOWING

Sow chives directly outdoors as soon as the soil begins to warm in spring. Draw out a row 1cm (½in) deep and sow seeds thinly along it. Cover them over with soil, water well, and label the row. Space the rows 30cm (12in) apart.

If raising chives to plant among other species, it's best to sow seeds indoors in small pots in early spring and plant out later, as tiny seedlings can easily be crowded out. Sow on to the compost surface, cover lightly with compost, water, and place on a warm windowsill. Young chive plants are also available from garden centres and nurseries.

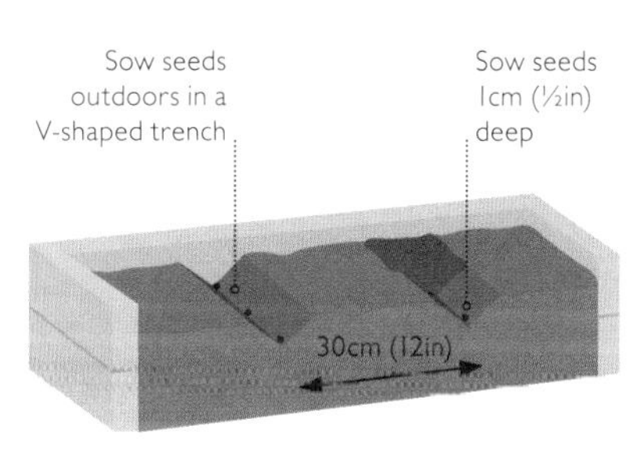

Plant out indoor-raised chives in spring after acclimatizing them to the outdoors.

GROW

Gradually thin out seedlings growing outdoors to 23cm (9in) apart; don't waste the thinnings, which make a tasty addition to salads. Water plants regularly in summer to stimulate new growth and feed those in containers with a balanced liquid fertilizer weekly. Remove faded flowers to prevent chives seeding all over the garden. To retain vigour, lift and divide established clumps every three years and replant smaller portions with healthy roots and shoots.

Chives are compact and the whole plant can be eaten.

HARVEST

Leaves can be cut as required with scissors about eight weeks after sowing. The flowers are also edible, and have a mild onion flavour.

VARIETIES

Species chives (*Allium schoenoprasum*) are most common, but seek out other varieties for colour and flavour.

'CHA CHA' Identical to the species, apart from the spiky tufts of vibrant green leaves produced instead of flowers.

'FORESCATE' A vigorous, strong-flavoured variety with blue-tinged leaves and attractive pink flowers.

GARLIC CHIVES This is a different species, which tastes distinctly of garlic.

LEEKS

The sweet, white stems of leeks stand ready to harvest through winter frost and snow, making them the ideal basis for hearty soups and stews. Their long growing season can make them tricky to squeeze into a small space, so plan to plant leeks to follow early summer crops.

DIFFICULTY Easy
WHEN TO SOW Feb to Mar (indoors); Apr (outdoors)
IDEAL SOIL TYPE Fertile and well-drained
SITE REQUIREMENTS Full sun
GERMINATION TIME 10–14 days
GROW FROM Seeds or plants
YIELD 5kg (11lb) from a 2m (6ft) row

CALENDAR

	WINTER			SPRING			SUMMER			AUTUMN		
SOW			Sown indoors	Sown indoors	Sown outdoors							
HARVEST	■	■	■	■					■	■	■	■

Sown indoors
Sown outdoors

Time between sowing and harvesting
5–7 months

SOWING

Leeks are rarely sown into their final location because they take such a long time to grow from seed. Traditionally, they are sown in a seedbed outdoors in spring, to produce young leeks to transplant into the vegetable patch in early summer. Where there isn't space for this, leeks do just as well started in containers indoors. Try sowing two or three different varieties in succession for continuous crops from late summer to early spring. Sow early varieties from late winter to early spring, mid-season varieties in early spring, and late varieties in mid-spring.

Fill a tray with compost to 2cm (1in) below its rim

Allow 2cm (1in) between seeds

Cover with 1cm (½in) of compost

SOW IN CONTAINERS Fill a deep seed tray, module tray, or pot about 17cm (6½in) in diameter with multi-purpose compost to 2cm (1in) from the top. In trays and pots, sow seeds thinly across the compost, with about 2cm (1in) between them. Sow a single seed into each module. Cover with 1cm (½in) of compost, water well, and label. Place late winter and early spring sowings on a cool windowsill at 10–15°C (50–60°F); mid-spring sowings can be kept on the windowsill or in a sunny spot outdoors.

SOW INTO SOIL Wait until the soil has warmed in mid-spring before sowing leeks outdoors. Give them fertile soil in full sun and rake it to a fine crumb before sowing. Draw out a row 1cm (½in) deep along a string line and sow seeds thinly along it so they are 2cm (1in) apart. Pull soil back over the row, water thoroughly, and label.

Sowing leek seeds into individual modules minimizes disruption to the root system when they are later planted out.

GROW

TRANSPLANT Plant young leeks out in early summer when they reach a height of about 20cm (8in). They are best planted in fertile soil and full sun, but can also be grown in large, deep containers.

Separate young plants that you have raised in seedbeds, pots, or trays by gently teasing apart their roots; drop each individual plant into a 10–15cm- (4–6in-) deep hole made with a dibber or a broom handle. Planting the leeks at this depth helps to produce blanched white stems. Leave the soil in place around module-raised leeks and plant at the same depth. Space plants at 10–15cm (4–6in) intervals in rows 30cm (12in) apart.

WATER Rather than backfilling each hole with soil after transplanting, fill it with water using the spout of a watering can; soil will gradually settle around the plant. Water daily while plants establish and during dry weather. Feed with a balanced liquid fertilizer fortnightly during summer.

WEED Leeks have slender foliage that does little to shade out weeds, so weed regularly among the rows with a hoe or hand fork.

Leeks are nutrient-hungry and so require fertile soil for success.

HARVEST

Leeks can be harvested as required from late summer, either by pulling the stem in soft ground or with the help of a fork on firmer soil. Most varieties will wait in the soil for at least three months until you need them and late varieties will stand through freezing winter weather. Lift and use the last of your crop before they bolt and send up flower stems.

Leeks are ready to harvest when their stalks are approximately 2.5cm (1in) across.

VARIETIES

Choose early varieties for a faster crop, but for hardy leeks to stand in the soil through winter, select a mid- or late-season variety.

'CRUSADER' A hardy late variety that will stand ready to harvest from November until spring.

'MUSSELBURGH' This reliable mid-season variety crops through the winter.

'NORTHERN LIGHTS' The leaves of this late variety develop a purple tinge in cold winter weather.

'OARSMAN' A vigorous, mid-season variety that bulks up fast to harvest from November.

'PANCHO' An early variety with good resistance to bolting, this plant is ready to harvest from late summer until Christmas.

Hardy leeks will endure winter frost and snow, ready to be lifted as needed.

TROUBLESHOOTING

Perhaps thanks to their strong odour, plants of the onion family are not troubled by many pests, but those insects that have developed a taste for them can cause serious damage. Problems with fungal diseases are by far the most serious threat, but can be minimized by growing in well-drained soil and practising crop rotation.

ALLIUM LEAF MINER

PROBLEM Lines of white spots on foliage; small white maggots or brown pupae found in stems and bulbs; plants may start to rot.
CAUSE Adult flies sucking sap from leaves; maggots feeding inside leaves, stems, and bulbs.
REMEDY Cover with fine mesh during spring and autumn to deter the flies.

BIRDS

PROBLEM Onion and shallot sets are pulled out of the soil after planting.
CAUSE Birds, often blackbirds, unearthing sets while feeding.
REMEDY Uprooted sets are not usually damaged and can be replanted. Where the problem is persistent, cover the crop with netting or fleece until green shoots appear from the sets.

LEEK MOTH

PROBLEM Foliage of leeks, onions, and garlic develops pale patches; tunnels appear in stems and plants may rot.
CAUSE The small, creamy-white caterpillars of the leek moth.
REMEDY Protect with fleece or fine mesh from late spring until late autumn; squash cocoons found on foliage; rotate alliums around the garden each year.

ONION FLY

PROBLEM Onion, shallot, and leek seedlings die, and foliage of larger plants yellows and wilts.
CAUSE The maggots of the onion fly feeding on roots and bulbs.
REMEDY Grow crops under mesh to exclude egg-laying flies. Use a nematode preparation to control maggots; grow from sets rather than seed.

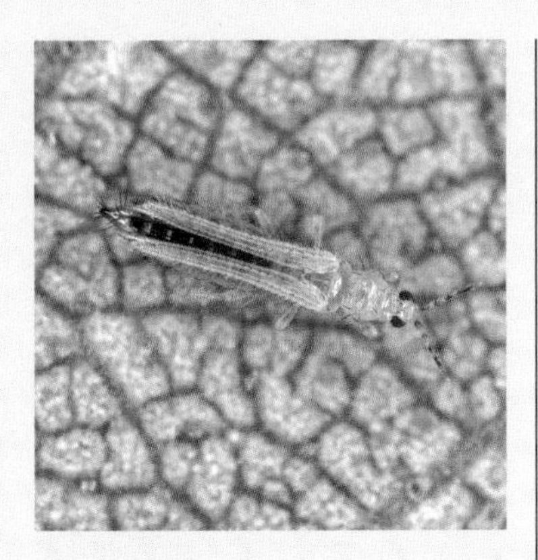

ONION THRIPS

PROBLEM Pale mottling appearing on the foliage of leeks, onions, shallots, and garlic.
CAUSE Small, sap-sucking insects that grow to about 2mm ($^1/_{12}$in) long.
REMEDY Healthy plants can usually tolerate damage. Organic pesticide sprays, such as pyrethrum, are effective, but always read the maker's guidelines.

BOLTING

PROBLEM Flower stalks are produced before crops are mature.
CAUSE Temperature fluctuations, particularly during germination and early growth, and hot, dry weather.
REMEDY Protect sowings from cold by growing indoors or under fleece. Water during dry weather. Immediately harvest crops that begin to bolt.

DAMPING OFF

PROBLEM Seedlings collapse and die, and white mouldy growth appears.
CAUSE Fungi infecting seedlings.
REMEDY Always use clean pots and fresh compost. Sow seeds as thinly as possible and provide good ventilation in propagators as soon as the seedlings germinate to reduce the humidity around plants.

LEEK RUST

PROBLEM Raised, bright orange spots appear on the foliage and stems of allium crops.
CAUSE A fungal disease.
REMEDY Give plants plenty of space to increase air circulation. There is no cure; remove badly infected plants, rotate alliums around your plot, and grow resistant varieties.

ONION DOWNY MILDEW

PROBLEM Leaves of onions, shallots, spring onions, and chives turn yellow; this is followed by white and sometimes purple fungal growth.
CAUSE A fungus-like organism.
REMEDY Dispose of affected plants; do not compost them. Maximize air circulation by spacing plants widely and removing weeds. Practise crop rotation.

ONION WHITE ROT

PROBLEM Leaves wilt and turn yellow, and the bases of bulbs and roots rot, showing fluffy white growth.
CAUSE A soil-borne fungal disease.
REMEDY There is no cure and the fungus remains in the soil for many years. Avoid transferring soil to your garden from infected plots; plant only healthy onion sets and garlic.

Tomatoes and peppers are popular warm-weather crops that have similar requirements for growth and so are often planted in the same part of the garden.

SUMMER VEGETABLES

Fast-growing, free-flowering, and laden with colourful fruit, these showstoppers are fun to grow. They make perfect dishes to enjoy outdoors with friends and are delicious tossed into salads, chargrilled, or chopped into a spicy salsa.

SUMMER TREATS

The huge range of summer veg varieties that you can grow from seed bears little resemblance to the limited selection available in the shops. Heritage tomatoes, exotic chillies, and gourmet squashes should fire your enthusiasm for growing your own. Seeds can be sown indoors in early spring, or even late winter, for a vibrant bounty from midsummer well into autumn. The aroma and intense flavour of homegrown tomatoes, the sugary crunch of just-picked sweetcorn, crisp, juicy cucumbers, and buttery baby courgettes will have you hooked from the first bite.

TURN UP THE HEAT

Summer vegetables need heat and a long growing season to ripen a good crop, which means that in temperate climates they all benefit from the head start provided by sowing on a warm windowsill indoors. It's then essential to squeeze as much warmth and sunlight as possible out of summer by planting them out into a really sunny, sheltered spot outdoors, or a greenhouse if you have one. Compact or bush varieties are the best bet for cooler regions, as they grow quickly and usually produce small fruits which ripen readily. Fuel summer growth with regular watering and feeding, and keep it consistent to prevent fruits being spoiled or splitting.

A SUITABLE SPACE

Bush tomatoes, chillies, peppers, and aubergines all flourish in containers, and make a decorative addition to a sunny patio or balcony when dripping with fruits. The portable nature of pots also allows plants to be moved indoors to continue cropping when the weather cools in autumn. Where space in beds is limited, there is also success to be had growing tall cordon tomatoes, cucumbers, and courgettes in large containers or grow bags. Vigorous winter squashes and perennial globe artichokes need to sink their roots deep into open soil, however, while sweetcorn is best planted in a block pattern in a bed to promote successful pollination.

BUSH TOMATOES

Give these tender, heat-loving plants a sunny spot outdoors and they will repay you with generous crops of irresistible cherry tomatoes, no fiddly staking or side-shooting required. Most varieties spread 60–90cm (2–3ft) and look great cascading from containers or hanging baskets.

DIFFICULTY Easy
WHEN TO SOW Mar (indoor crop); Mar to Apr (outdoor crop)
IDEAL SOIL TYPE Fertile, well-drained, with a pH of over 6.5
SITE REQUIREMENTS Full sun; sheltered
GERMINATION TIME 3–14 days
GROW FROM Seeds or plants
YIELD 1.8–2.5kg (4–5½lb) per plant

CALENDAR

	WINTER	SPRING	SUMMER	AUTUMN
SOW				
HARVEST				

Indoor crop
Outdoor crop

Time between sowing and harvesting
3–4 months

SOWING

Fill modules or pots with compost, and make a dent about 5mm (¼in) deep in the centre of each. Drop one seed into each dent, cover with compost, water well, and allow to drain. Place the pot in a propagator with a clear lid or cover it loosely with a clear plastic bag: both act like a mini-greenhouse, keeping in heat and moisture. Place on a warm, bright windowsill early in spring and plant outdoors as soon as the risk of frost has passed. Alternatively, the plants can be kept indoors for an earlier crop.

When seedlings emerge, remove the lid or bag to reduce humidity. Water regularly and turn pots daily to prevent seedlings bending towards the light.

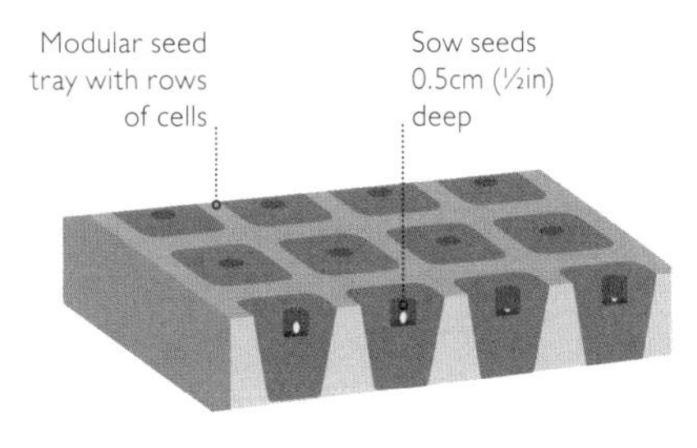

Sow seeds into moist compost. You should have sturdy seedlings within a few weeks.

BUYING PLANTS If you have no space to raise tomatoes indoors, or you only want one or two plants, then buy young plants from a garden centre. Choose stocky, dark green specimens and gradually introduce them to outdoor conditions before planting out.

GROW

POTTING ON Repot young plants into progressively larger pots as they grow, and space them out as much as possible on the windowsill. In late spring, begin acclimatizing the plants to outdoor conditions by moving them outside for longer periods each day; after two weeks they can be left out overnight.

PLANTING OUT Choose a container at least 23cm (9in) in diameter, and fill it to about 5cm (2in) from the top with compost. Alternatively cut two planting holes in the top of a growbag or prepare a bed of soil, enriched with well-rotted compost or manure. Once the risk of frost is over, plant the young tomatoes in your chosen spot slightly deeper than they were growing in their original pots; water in thoroughly.

When potting on, avoid holding young plants by the stem, which can bruise easily.

WATER At first, water plants in pots regularly, but avoid overwatering. Plants in the ground initially need watering only during dry weather. Once the flowers form, water regularly to keep the roots consistently moist or install a drip irrigation system (see *p.37*). Container plants will need watering daily.

FEED Use a high potash fertilizer, such as liquid tomato feed, at alternate waterings once the first fruits form.

PROTECT Cover plants with fleece if a cold night is forecast. Manually remove slugs and snails from around the plants.

Drip irrigation systems efficiently deliver water directly to the base of the plant.

HARVEST

Vine-ripened tomatoes have the sweetest taste; pick fruits as they ripen and eat as soon as possible. Never store tomatoes in the fridge as it destroys the intensity of their flavour. As the weather cools in autumn pick all remaining fruit and ripen in a bowl indoors. Avoid placing on a sunny windowsill to ripen: they may rot.

When picking, pull tomatoes gently by holding the stem with one hand and the fruit with the other, breaking the stalk just above the calyx.

VARIETIES

Bush tomatoes tend to bear small but highly flavoured fruit. Choose a variety with some blight resistance if your area has wet summers.

'KORALIK' An early variety from Eastern Europe that performs well in cool, wet summers and has good blight tolerance.

'LIZZANO' A vigorous trailing variety for containers or the veg plot, producing heavy trusses of red cherry tomatoes.

'LOSETTO' Abundant small fruits are quick to ripen and are packed with sweetness.

'RED ALERT' Fast-maturing and heavy yielding: ideal for outdoor cultivation.

'TUMBLER' A classic compact, trailing bush variety, perfect for hanging baskets and containers.

WHY NOT TRY?

Even if space is really tight you can always make room for dwarf tomatoes on a sunny patio or a wide windowsill. Dwarf bush varieties, such as 'Balconi Red' and 'Micro Tom', are incredibly compact, with a height and spread of 30cm (12in) or less, and can be grown in smaller pots than larger varieties: with careful feeding and watering they can thrive in a container that is just 12cm (5in) in diameter. The smaller size of the plants means that the yield of cherry-sized fruit is modest, but they are perfect for supplying sweet, home grown summer snacks.

Dwarf varieties such as this 'Patio Plum' can be surprisingly productive, but most have a short growing season.

CORDON TOMATOES

Cordon tomatoes are tall, heat-loving plants with a single main stem that need to be grown up strong supports. They generally have a better flavour and heavier yield than bush tomatoes, but may take longer to ripen. Grow them outside in warm regions and under cover where summers are cool.

DIFFICULTY Easy
WHEN TO SOW Mar (unheated greenhouse); Apr (outdoors)
IDEAL SOIL TYPE Fertile, well-drained, with a pH over 6.5
SITE REQUIREMENTS Full sun and sheltered
GERMINATION TIME 3–14 days
GROW FROM Seeds or plants
YIELD 2–4kg (4.5–9lb) per plant

CALENDAR

	WINTER			SPRING			SUMMER			AUTUMN		
SOW				Unheated greenhouse	Outdoors							
HARVEST								Unheated greenhouse	Unheated greenhouse / Outdoors	Unheated greenhouse / Outdoors	Unheated greenhouse	

Unheated greenhouse
Outdoors

Time between sowing and harvesting
4 months

SOWING

Tomatoes need a long growing season and cannot tolerate frost. Give them a head start by sowing indoors from early spring to plant into a greenhouse in late spring, or to plant outside in early summer. Don't sow too early as plants will fast outgrow space on a windowsill.

Start off tomatoes in modules kept in a warm, bright location.

SOW IN MODULES Fill a module tray with multi-purpose compost and sow a single seed 2cm (1in) deep into each module. Cover with compost, label, and water well. Enclose the tray in a clear plastic bag or place into a propagator; position on a sunny windowsill to provide extra warmth. Once the seedlings have germinated, remove the covers and water to keep the compost moist. Pot each plant up into a 9cm (3½in) pot when it reaches about 8cm (3in) in height and has three true leaves (after the first two "seed leaves"). Space out the plants so they do not touch.

BUYING PLANTS If you have no space to raise tomatoes indoors, or you only want one or two plants, then buying young plants from a garden centre or by mail order is the best option. Choose stocky, dark green specimens.

GROW

PLANT OUT Acclimatize plants raised under cover to outdoor conditions by moving them outside for progressively longer each day over two weeks in late spring. Plant out in early summer, once the risk of frost has passed, or into an unheated greenhouse in late spring.

Cordon tomatoes thrive in full sun, when planted into fertile soil, growbags, or containers at least 23cm (9in) in diameter. Space them about 45cm (18in) apart and plant them deeply with their lowest leaves just below soil level. Add strong supports, such as canes or metal spirals, at least 1.5m (5ft) tall for each plant and tie in the stem loosely with garden twine.

Plant out and tie in young plants. Be careful not to bruise their soft stems.

TRAIN UP SUPPORTS Keep the main stem of each plant tied to its support as it grows. Pinch out side shoots, which form between each leaf and the main stem, to stop bushy growth and encourage flowering. Pinch out the growing point at the tip of each plant once about five trusses of fruit have set, to divert resources to ripening.

Pinch out side shoots as soon as they appear on the main stem.

WATER AND FEED Outdoor tomatoes need watering only in very dry weather before flowering, and soaking weekly if conditions remain dry after that. In fertile soil they need little feeding, but benefit from a weekly feed with a liquid tomato fertilizer as fruit trusses form.

Plants grown in containers or under cover need watering once or even twice a day in summer to maintain healthy growth. Once the first truss of fruit sets, feed with liquid tomato fertilizer twice a week.

Water container plants until water runs freely from the bottom.

HARVEST

Pick fruits as soon as they ripen by snapping their short stems where they join the vine. Tomatoes are spoiled by cold conditions, so make sure to harvest outdoor crops before the first frost and never store picked tomatoes in the fridge. Unripe fruit can be cut from the plant on the vine and brought indoors to ripen. Green tomatoes can also be made into chutney.

Allow tomatoes to ripen fully and develop a rich colour before picking to enjoy them at their sweetest.

VARIETIES

Those who think tomatoes only come in red will get a shock when they see the range of shapes, sizes, and colours available. Sampling the flavours of a range of varieties is a huge perk of growing your own.

'FERLINE' Good blight tolerance and high yields of large, tasty tomatoes make this an ideal outdoor variety.

'GARDENER'S DELIGHT' A trusted variety with large cherry fruits; it performs well outdoors and under cover.

'SUNGOLD' The orange-coloured, cherry fruits of this variety are among the sweetest of all tomatoes.

'SWEET MILLION' Cascading trusses of small, sweet tomatoes ripen well outdoors.

'SWEET OLIVE' This flavoursome baby plum variety produces bumper crops.

'TIGERELLA' Delicious orange-red fruits retain attractive green stripes when ripe.

COURGETTES AND SQUASHES

These large, fast-growing plants are hugely productive. Summer squashes, including courgettes and patty pan types, tend to have a bushy habit and are eaten young, while trailing winter squashes mature on the plant until autumn.

DIFFICULTY Easy
WHEN TO SOW Apr–May (indoors); June (outdoors)
IDEAL SOIL TYPE Fertile, well-drained
SITE REQUIREMENTS Full sun and sheltered
GERMINATION TIME 3–7 days
GROW FROM Seeds or plants
YIELD Up to 20 fruits per plant

CALENDAR

	WINTER	SPRING	SUMMER	AUTUMN
SOW				
HARVEST				

Summer squash
Winter squash

Time between sowing and harvesting
2–3 months

SOWING

Courgettes and squashes are tender plants which need heat to germinate, so are best sown indoors in mid- to late spring, about a month before the last frost is expected. Sow single seeds 2cm (1in) deep into modules or small pots filled with multi-purpose compost, water well, label, and cover with a clear plastic bag or a lidded propagator and place on a warm windowsill.

Direct outdoor sowing into soil improved with plenty of organic matter is possible in early summer. Place two seeds into one hole 5cm (2in) deep at stations 90cm (3ft) apart. Cover with soil, water, then protect under a cloche for additional heat. If you only need a few plants or there is no space to raise them indoors, courgettes and squashes can also be bought as young plants from garden centres or mail order.

Courgettes are very vigorous, so don't sow too many seeds.

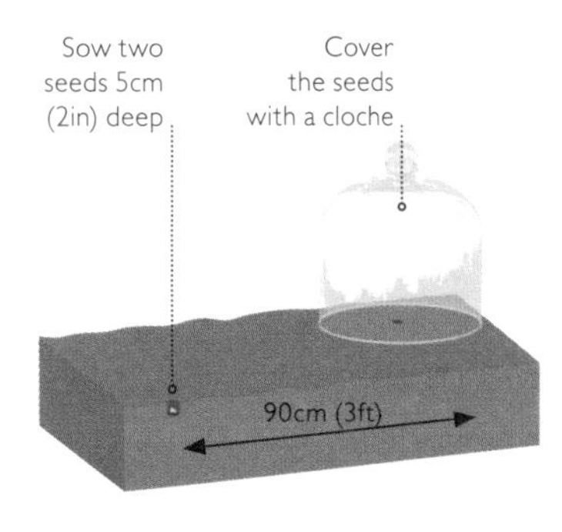

GROW

PLANT OUT Before transplanting, prepare the soil for these hungry plants by incorporating plenty of well-rotted manure or compost. Create a mound of soil about 15cm (6in) high on heavy soils to aid drainage. Harden off plants by acclimatizing them to outdoor conditions over 2–3 weeks. Courgettes and squashes can also be grown in large containers or growbags.

Space plants 90cm (3ft) apart; place them in the soil at the same depth as they were growing in their pots, firm the soil gently around their roots, and water. Cover with cloches to protect their delicate stems from breezy conditions.

Avoid disturbing the roots when planting out squashes and courgettes.

FEED AND WATER These soft plants are prone to stem rots, so only water them in dry weather or if growing in sandy soil. Plants grown in pots, however, will need frequent watering – daily or even twice daily in summer – to support their rapid growth. Feed container-grown plants with a liquid tomato fertilizer twice a week after their first fruits have formed. Plants growing in the open soil only need feeding if they are not growing vigorously.

Pressing the male flower on to the female flower results in pollination.

POLLINATE Courgettes and squashes produce separate male and female flowers, and insect pollination needs to occur for fruits to form naturally. Yields can be improved, particularly in cold or wet spells of weather, by hand pollinating the female flowers, which have a small, immature fruit below them. Do this by picking an open male flower with a long, thin stem, removing its yellow petals and pushing it into the centre of a female flower.

For large winter squashes, limit each plant to three fruits by pinching out the growing tips of plants and removing all unwanted flowers.

HARVEST

Courgettes and summer squashes taste best when young. Harvest by twisting or cutting from the plant. Picking when young encourages the production of more flowers and fruits, and avoids having to deal with gigantic watery marrows later in the season.

Winter squashes should be left on the plant to develop. At the end of summer, remove any leaves shading the fruit and allow their skin to harden in the sun. Harvest before the first frost by cutting, leaving a length of stem attached. Store in a dry, frost-free place and they will keep for several months.

Squash flowers are edible and are delicious stuffed with ricotta cheese.

VARIETIES

Most courgette varieties taste quite similar but have varying growth habits. Choose winter squash by fruit size and storability after harvest.

'DEFENDER' A prolific, disease-resistant courgette with tasty green fruit.

'HUNTER' This tasty winter squash, with rich orange flesh, is bred to mature early in cooler climates.

'MIDNIGHT' This compact green-speckled courgette is perfect for containers.

'SUNBURST' This patty pan summer squash produces yellow UFO-like fruit.

'UCHIKI KURI' This onion-shaped winter squash matures fast and has delicious flesh.

WHY NOT TRY?

Many squash varieties and climbing courgettes can be trained up a trellis or arch – a great way to fit these vigorous plants into small spaces. Their huge leaves quickly crowd out plants at ground level, but tying the main stem on to a sturdy support directs growth upwards, leaving plenty of space below for other crops. Their bright flowers and colourful fruits make an attractive summer feature. The climbing courgette varieties 'Black Forest' and 'Shooting Star' are well suited to vertical growing. The bizarrely shaped squash 'Tromboncino' and most winter squashes are also perfect for training up supports.

Colourful pendent fruits make squashes decorative climbers.

RIDGE CUCUMBERS

Cucumbers can be grown outdoors if you choose the right variety. Ridge cucumbers produce shorter, rougher-skinned fruits than the standard greenhouse cucumbers and, though they will not survive frost, they are tougher and more resistant to pests, diseases, and cooler outdoor conditions.

DIFFICULTY Easy
WHEN TO SOW Apr–May (indoors); June (outdoors)
IDEAL SOIL TYPE Fertile, well-drained
SITE REQUIREMENTS Full sun and sheltered
GERMINATION TIME 7–10 days
GROW FROM Seeds or plants
YIELD 20–30 fruits per plant

CALENDAR

	WINTER			SPRING			SUMMER			AUTUMN		
SOW					Indoors	Indoors	Outdoors					
HARVEST								Indoors / Outdoors	Indoors / Outdoors	Indoors / Outdoors	Indoors / Outdoors	

Indoors
Outdoors

Time between sowing and harvesting
3 months

SOWING

For best results, sow ridge cucumbers into small pots indoors in mid- to late spring. Dib a hole 2cm (1in) deep at the centre of each pot and drop in a single seed. Cover with compost, water well, label, and place in a clear plastic bag or propagator. Keep the pots in a warm room (or a propagator) at 20°C (68°F) to germinate.

Seeds can also be sown outdoors into warm soil in early summer. Add plenty of organic matter to the soil beforehand and sow two seeds 2cm (1in) deep at each position (so allowing for the failure of one of the seeds). Water well and cover the planting with cloches. Remove the weaker seedling if both should germinate. Where space to raise plants indoors is limited, or if you need only a few plants, consider buying young plants from a garden centre or by mail order.

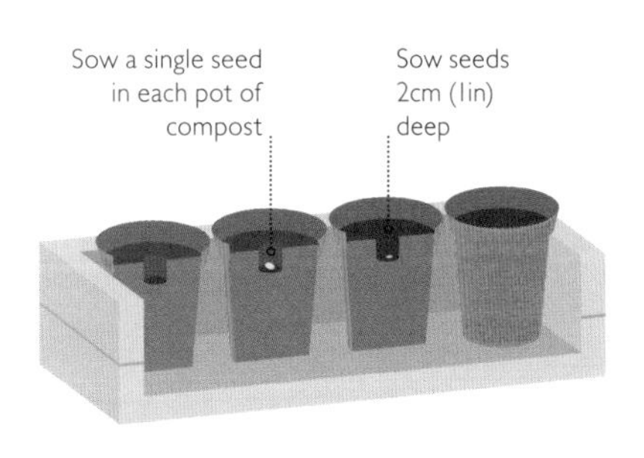

Don't sow too many seeds: four or five healthy plants will produce enough fruit for a family of four.

GROW

POT ON Seedlings grow rapidly and need to be transferred into 9cm (3½in) diameter pots 7–10 days after they have germinated. Water them well and continue to grow on a warm windowsill at about 15°C (51°F).

SUPPORT Cucumbers can be grown along the ground if planted 90cm (3ft) apart, but can be planted closer together (45cm/18in apart) if trained up a trellis, cane tripods, or netting to a height of about 1.8m (6ft). Supports are best put in place before planting. Tie the main stems on to the supports regularly as they grow.

A cucumber support need not be a traditional wooden trellis or pyramid.

PLANT OUT Harden off plants over two weeks by moving pots outdoors for increasing periods. Once the risk of frost has passed, plant in fertile soil in full sun in a sheltered spot. On heavier soils, plant on mounds 15cm (6in) high for better drainage. Place the plant into a hole the same depth as the pot, firm soil around the roots, and water well. Cover with cloches. Cucumbers can also be planted into large containers or growbags.

WATERING AND FEEDING Water plants growing in the soil weekly and those planted in containers once or twice a day during hot weather. Cucumbers are hungry crops and even those planted directly in the soil will benefit from a regular liquid feed with tomato fertilizer once fruits start to form; plants in containers should be fed twice a week.

POLLINATION Ridge cucumbers have separate male and female flowers and therefore require insect pollination in order to produce fruit. Some "all female" varieties are also grown outdoors and will produce bitter cucumbers if pollinated by ridge varieties: don't grow them together.

Three cucumber plants will have space to thrive in one large growbag.

Water well to bulk the fruit, but don't overwater as this may cause stems to rot.

HARVEST

Pick cucumbers as soon as they are a usable size by cutting the stem with a sharp knife or secateurs. Try to harvest them in the morning when fruits are cool and crisp. They are prolific plants in warm summers, so make pickles for use in winter with any excess fruit.

Harvest when the fruits reach a length of about 20cm (8in).

WHY NOT TRY?

Cucamelons have crisp, juicy flesh that tastes like cucumber with a splash of lime. They are delicious for snacking or adding to salads and can be grown in the same way as ridge cucumbers in containers, in soil, and trained up supports. Cucamelons thrive in heat, and are best sown indoors in mid- to late spring to be planted out in a sunny, sheltered spot once nights are no longer frosty. In hot weather the vines can reach 3m (10ft) tall, so should be pinched out at the top of their supports to encourage fruiting. When picked regularly, cucamelons will crop from mid-summer until early autumn.

The cucamelon resembles a miniature watermelon.

VARIETIES

Traditional ridge cucumbers have prickly skins, 'Burpless' types are longer and smoother, and heirloom varieties have pale-skinned, rounded fruits.

'BURPLESS TASTY GREEN' One of the best-tasting outdoor cucumbers, with long fruits that are crisp and almost spine-free.

'CRYSTAL APPLE' Produces yellow-skinned, rounded fruits that have a delicate, slightly sweet flavour.

'DIAMANT' Produces abundant small cucumbers ideal for pickling.

'MARKETMORE' An excellent, disease-resistant ridge cucumber, bearing tasty, spiny-skinned fruits 20cm (8in) long.

AUBERGINES

Aubergines are attractive, bushy plants with soft, grey-green leaves, purple flowers, and glossy fruits that range from deep purple to white. They look great in a pot on a sunny patio, but only flourish outdoors in warm areas. Grow them in a greenhouse or on a windowsill where summers are cool.

DIFFICULTY Moderate
WHEN TO SOW Feb to Mar (under cover)
IDEAL SOIL TYPE Fertile, well-drained
SITE REQUIREMENTS Sunny and sheltered
GERMINATION TIME 7–14 days
GROW FROM Seeds or plants
YIELD 5–10 fruits per plant

CALENDAR

	WINTER			SPRING			SUMMER			AUTUMN		
SOW			■	■								
HARVEST									■	■	■	

Time between sowing and harvesting
4–5 months

SOWING

In late winter or early spring, sow two seeds per small pot, 5mm (¼in) deep. Cover with compost, label, and water well. Keep the pots at about 20°C (68°F) in a propagator or in a clear plastic bag on a warm windowsill. Once germinated, keep on a bright windowsill above 15°C (59°F) and move them into 9cm (3½in) diameter pots when about 5cm (2in) tall.

GROW

PLANT OUT Harden off seedlings to be grown outdoors and only plant them out after all risk of frost has passed. Warm soil under fleece before planting. Plant the seedlings in a sunny, sheltered spot, spaced 45cm (18in) apart, or 30cm (12in) apart for dwarf varieties. Alternatively move the plants into 20cm (8in) diameter pots or growbags to grow outdoors, in a greenhouse, or next to a bright window indoors.

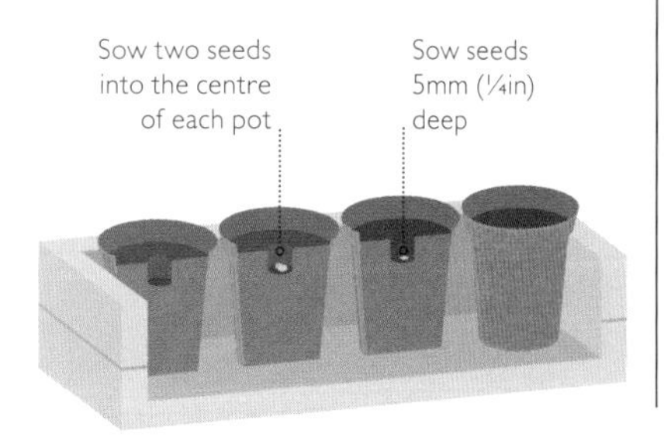

SUPPORT AND PINCH OUT Tie plants in to supporting canes as they grow. When they reach 25cm (10in) tall, pinch out the main growing tip to encourage side shoots. Remove flowers once five or six fruits have set.

Tie string between adjacent canes to support the bushy plants.

FEED AND WATER Water plants regularly in hot weather and those grown in pots once or twice daily. Mist the foliage every day after flowering begins to help set fruit. Feed weekly with a liquid tomato fertilizer once the first fruits set.

HARVEST

Pick aubergines from late summer once fruits are plump, smooth, and glossy, using secateurs to cut the stem.

Harvest aubergines grown outdoors well before the first frosts.

VARIETIES

'MONEYMAKER' Produces early outdoor crops of drop-shaped purple aubergines.

'POT BLACK' These compact plants are perfect for pots and yield abundant, early ripening, dark purple, miniature fruits.

SWEET AND CHILLI PEPPERS

Vibrantly coloured sweet peppers and spicy chillies are exciting crops to grow. It takes a hot summer to ripen them outdoors, but peppers thrive in a greenhouse in cooler areas and smaller, bushy varieties make fantastic houseplants.

DIFFICULTY Moderate
WHEN TO SOW Feb to early Apr (indoors)
IDEAL SOIL TYPE Fertile, well-drained
SITE REQUIREMENTS Sunny and sheltered
GERMINATION TIME 7–21 days
GROW FROM Seeds or plants
YIELD 5–50 fruits per plant depending on variety

CALENDAR

	WINTER	SPRING	SUMMER	AUTUMN
SOW				
HARVEST				

Time between sowing and harvesting
4–5 months

SOWING

Start early, in late winter or early spring, to provide the long growing season that peppers need to ripen. Sow two seeds 5mm (¼in) deep in small pots or modules, cover lightly with compost, water, and germinate at 21°C (70°F) in a heated propagator or on a warm windowsill. Grow on seedlings in warm conditions and transfer into 9cm (3½in) pots when about 5cm (2in) tall. Buy young plants if there is no space to raise seedlings or you forget to sow early.

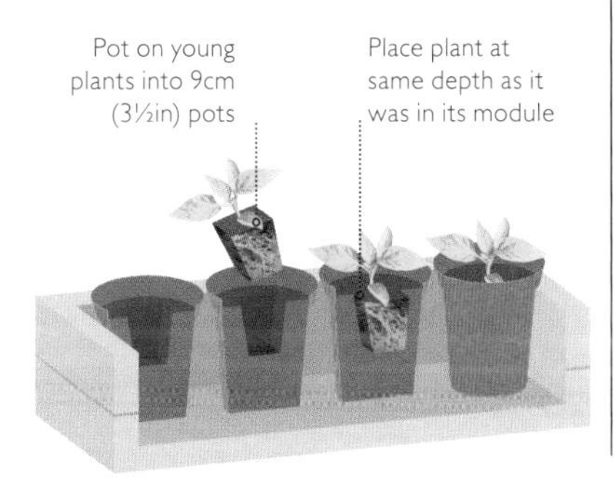

GROW

PLANT OUT Harden off plants gradually before planting them outdoors in early summer. Plant them directly into the ground 30–45cm (12–18in) apart or into containers or growbags. Firm the soil around the roots, water well, and protect outdoor plants with cloches on chilly nights.

FEED AND WATER Water regularly, but avoid overwatering as it degrades flavour. Feed plants in pots with a liquid tomato fertilizer every ten days once fruits set, but only feed those in the soil if growth is poor. Remove the lowest fruit to encourage vigorous, bushy growth. Mist plants under cover to help fruit set and reduce pest problems.

HARVEST

Peppers take a long time to ripen; pick the first two or three while still green to allow more to form and ripen later.

Harvest by cutting the stem of each fruit with secateurs.

VARIETIES

Plants vary greatly in height as well as in flavour; choose a variety to suit your taste and the space available.

'BASKET OF FIRE' These compact plants, just 30cm (12in) tall, are covered in tiny, hot chillies in cream, yellow, and red.

'HUNGARIAN HOT WAX' Medium-heat chillies are borne on tall, elegant plants.

'KING OF THE NORTH' Tall, vigorous plants produce stocky, thick-walled sweet peppers that ripen well outdoors.

'REDSKIN' This compact bell pepper is suited to pots and hanging baskets.

'THOR' These long, thin-walled, sweet "grilling" peppers ripen rapidly.

SWEETCORN

Fresh, fat sweetcorn kernels are the ultimate taste of late summer. Slender sweetcorn plants need open soil to flourish, but make an elegant feature in a border or a useful fast-growing screen where space is tight, and can easily be underplanted with another shorter crop.

DIFFICULTY Easy
WHEN TO SOW Apr–May (indoors); June (outdoors)
IDEAL SOIL TYPE Fertile, moist, well-drained
SITE REQUIREMENTS Full sun and sheltered
GERMINATION TIME 10–14 days
GROW FROM Seeds or plants
YIELD One or two cobs per plant

CALENDAR

	WINTER			SPRING			SUMMER			AUTUMN		
SOW					Indoors	Indoors	Outdoors					
HARVEST								Indoors / Outdoors	Indoors / Outdoors	Indoors / Outdoors	Indoors / Outdoors	

Indoors
Outdoors

Time between sowing and harvesting
3–4 months

SOWING

Traditional sweetcorn varieties are vigorous, but quickly lose their sweetness after picking. You may prefer to sow "Supersweet" types, which retain their sugars better, but need more moisture and heat to germinate, or "Tendersweet" varieties; these have persistent sweetness and thin-skinned kernels, but need warm conditions to perform well.

SOW INDOORS Give these heat-loving plants a head start while it's too cold to sow outside by sowing them indoors in mid- to late spring. Sow seeds singly, 2.5cm (1in) deep, in small pots, modules or cardboard tubes.

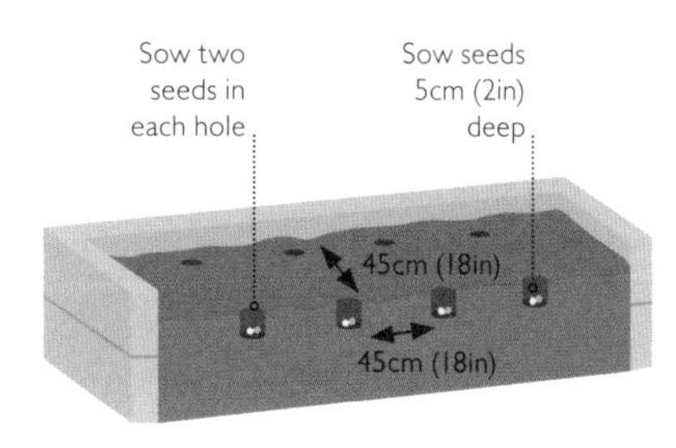

Cover them with compost, label, water them well, and place the containers on a warm windowsill or in a heated propagator. Don't sow too early if you live in a cool areas, because growth will be set back if young plants are kept in pots for too long. Buy young plants where there is no space to raise them from seed.

Sow sweetcorn seeds indoors individually in small containers.

SOW OUTDOORS Warm soil under fleece for two weeks before sowing outdoors in early summer. Cover seeds with soil, water well, and keep the soil moist during germination.

GROW

PLANT OUT Transplant sweetcorn outdoors in early summer, after the risk of frost has passed. Harden off plants by acclimatizing them to outdoor conditions gradually over two weeks. Plant in a sunny position, sheltered from strong winds. Space plants about 45cm (18in) apart each way, in blocks rather than long rows, to help achieve good wind pollination and produce cobs packed full of fat kernels. Water well after planting.

Try to minimize root disruption when planting out sweetcorn.

PROTECT In cold areas, cover the plants with fleece for a week or two after transplanting. Mice love sweetcorn, so set traps around outdoor sowings and developing cobs in areas where rodents may cause a problem. Create barriers around seedlings to keep slugs and snails at bay (*see p.38*). Mound earth around the bases of stems to help support the plants and tie them to canes if they start to lean.

WATER Only plants growing in very free-draining soil need regular watering, but all sweetcorn plants will benefit from watering during flowering. Add a mulch of compost around the plants once they are established to help retain soil moisture and suppress weeds.

Water at the base of plants as pollination occurs and cobs develop.

HARVEST

Check to see if cobs are ripe when their silky tassels turn brown. Pull back the protective leaves and pierce a kernel with your fingernail; if milky juice appears, the cob is ready to pick, but if it's clear, leave it on the plant a little longer to ripen. Twist each cob from the stem as soon as it is ripe and eat as soon as possible, before the sugars begin to turn into starch.

The sweet kernels only develop after successful pollination. Pollen is produced by the male flowers (the tassels on top of the stalk).

VARIETIES

Early varieties mature significantly faster than late varieties, allowing you to sow both at the same time for a continuous crop. Grow early varieties in cooler areas.

'EARLIBIRD' The earliest "Supersweet" variety gives a late summer harvest.

'LARK' This mid-season "Tendersweet" type produces cobs packed with sugary, thin-skinned kernels.

'RISING SUN' Tall, sturdy plants yield early crops of "Supersweet" cobs.

'SWEET NUGGET' Uniform cobs filled with large, golden kernels are ready to pick in early autumn from this mid-season "Supersweet" variety.

WHY NOT TRY?

Popping corn is a family favourite. Varieties specially bred for this purpose are grown in the same way as standard sweetcorn, but the cobs are dried, not eaten fresh. Leave cobs on the plant until their outer leaves turn yellow and brittle, and pick them in late autumn. In cool, wet areas (or where pests are likely to eat the kernels), harvest ripe cobs in early autumn, peel off the leaves, and hang them by their stalks to dry in a warm place, such as a sunny window or conservatory. When kernels are fully dry and start to fall from the cobs, empty them into an airtight container to store until needed.

Popping varieties include the red 'Strawberry'.

GLOBE ARTICHOKES

These imposing, silvery-leaved perennials work just as well in an ornamental border as they do in a veg plot. In summer, their scaly, globe-shaped buds are harvested as a delicacy. They are quick to establish from potted plants or seed and simple to propagate from offsets.

DIFFICULTY Easy
WHEN TO SOW Feb–Mar (indoors); Apr–May (outdoors)
IDEAL SOIL TYPE Fertile, well-drained
SITE REQUIREMENTS Full sun, sheltered
GERMINATION TIME 14–21 days
GROW FROM Seeds or plants
YIELD 10–12 heads per established plant

CALENDAR

	WINTER			SPRING			SUMMER			AUTUMN		
SOW			Indoors	Indoors	Outdoors	Outdoors						
HARVEST								Indoors/Outdoors	Indoors/Outdoors	Indoors/Outdoors		

Indoors
Outdoors

Time between sowing and harvesting
6 months (indoor sowings)

SOWING

Globe artichokes are easy to grow from seed, but the plants produced can be quite variable; it's a good idea to plant out more than you need with a view to keeping only the best specimens.

SOW INDOORS If you want a harvest in the first year, you will need to sow indoors in late winter or early spring. Fill modules with compost and sow the seeds singly 1cm (½in) deep, or fill a seed tray to 2cm (1in) below the top and sow seeds thinly over the surface. Cover with 1cm (½in) of compost, water well, label, and place on a warm, sunny windowsill.

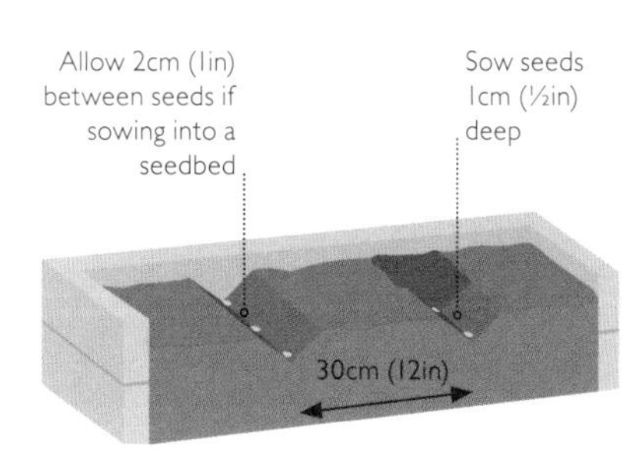

Sow artichoke seeds in a tray or modules. If you only need a few plants, it may be more convenient to buy plugs.

SOW OUTDOORS Wait until the soil has warmed in mid- to late spring before sowing outdoors. Raise young plants to transplant in summer by drawing out a row 1cm (½in) deep in a seedbed and sowing seed thinly along it. If you have space for a row of globe artichokes, sow three seeds every 90cm (3ft) along the row. Cover the seeds with soil, label, and water well.

GROW

PLANT OUT Harden off plants raised indoors or bought in pots by gradually introducing them to outdoor conditions. Choose a sheltered site in full sun and add plenty of well-rotted compost or manure to the soil before planting. In late spring or early summer, once the risk of frost has passed, plant out 90–120cm (3–4ft) apart, at the same depth as the plants were growing in their pots.

WATER AND FEED Water the young plants regularly during their first summer. Further watering should be unnecessary once these deep-rooted plants are established. Mulch around their bases with well-rotted compost in spring to preserve soil moisture, keep down weeds, and add extra nutrients.

Mulching around the plants helps keep the soil moist in summer.

PROTECT Watch out for blackfly on the flower heads – squash any that you see. Globe artichokes are hardy, but can be damaged by very cold winter weather, so cover them with fleece or a dry mulch of leaves or straw in late autumn.

Blackfly infestations can often be washed off using a garden hose.

PROPAGATE After 4–5 years, plants often lose vigour and are best lifted and replaced. Luckily, each clump forms small offsets at its edge, which can be sliced off with a sharp spade in spring. Replant only those with plenty of roots attached.

HARVEST

Globe artichokes will send up a single flowerhead late in the summer of their first year, but established plants should have heads to cut in midsummer, followed by a secondary crop in early autumn. Cut each stem with a sharp knife or secateurs, from when the head is tightly closed and the size of a golf ball to when the scales begin to open; if left any longer they become inedible.

Cut a 5cm (2in) length of stem when you harvest your artichokes; the stems of young buds are edible and make handling prickly varieties easier.

WHY NOT TRY?

Jerusalem artichokes are unrelated perennials grown for their delicious tubers. Unfussy about soil, and happy in shade, they are perfect for corners where little else will grow. They are also tall enough to create a screen or windbreak where required. Plant tubers 15cm (6in) deep, spaced 30cm (12in) apart, from late winter to mid-spring. Water during dry weather and pinch out any flower buds and the tops of plants when they reach about 2m (6ft) tall. Cut the stems down in late autumn and unearth the tubers as required.

Dig up all the tubers at harvest, or the plants will spread rapidly.

VARIETIES

The range of globe artichoke varieties is quite limited; the widest selection is available if you are willing to grow them from seed.

'GREEN GLOBE' This reliable green variety is widely available as seeds and young plants.

'ROMANESCO' A slightly later-maturing Italian variety, this is prized for its flavour and attractive purple flower heads.

'VIOLET DE PROVENCE' A striking purple French variety which sends up flower stems early.

'VIOLETTA DI CHIOGGIA' Each green scale on the head is flushed violet at its base, creating a highly decorative effect.

TROUBLESHOOTING

Difficult growing conditions cause more problems for summer veg than pests and diseases, so keeping plants warm, fed, and watered is the main route to plant health. Insect pests are more prevalent in sheltered greenhouses, though fungal diseases, such as blight, may become a problem, particularly during wet summers.

APHIDS

PROBLEM Clusters of green or black sap-sucking insects on young growth.
CAUSE Aphids, also known as greenfly and blackfly.
REMEDY Squash insects with your fingers as soon as they appear. Encourage aphid predators, such as ladybirds and hoverflies, into your garden with flowering plants.

RED SPIDER MITE

PROBLEM A pattern of fine, pale mottling on leaves grown under cover. Affected foliage may dry up and fall.
CAUSE Tiny, sap-sucking yellow-green mites on the undersides of leaves.
REMEDY Mites prefer dry conditions, so mist plants regularly in summer. Introduce a suitable biological control into an infected greenhouse.

WHITEFLY

PROBLEM Clouds of white insects fly up when plants growing under cover are disturbed; plant surfaces are sticky.
CAUSE Insects, known as whitefly, which secrete sticky honeydew.
REMEDY Hang sticky yellow traps to catch flying insects; spray plants with a suitable organic pesticide (check to ensure approval for use on edibles).

GREY MOULD

PROBLEM Fuzzy, light grey fungal growth appears on foliage, flowers, or fruits, particularly if they are damaged. Infected tissues turn brown and rot.
CAUSE A fungal disease called *Botrytis*.
REMEDY Remove any damaged parts. The fungus thrives in humid conditions, so avoid overcrowding plants and encourage good air circulation.

POWDERY MILDEW

PROBLEM A dusty, white covering spreads across courgette and cucumber leaves, particularly in late summer.
CAUSE A fungal disease.
REMEDY Remove infected leaves. Keep plants consistently watered and apply a mulch to slow the drying of the soil. Make space for air circulation between plants; choose resistant varieties.

TOMATO BLIGHT

PROBLEM Tomato foliage rots, collapses, and turns brown; fruits develop brown patches and decay.
CAUSE A fungus-like organism, which also causes potato blight.
REMEDY Dispose of infected plants in domestic waste. Grow resistant varieties or grow tomatoes under cover where infection is less likely.

TOMATO GHOST SPOT

PROBLEM Small, pale, circular marks on ripening tomatoes.
CAUSE Unsuccessful fungal infection.
REMEDY Fruits are marked but are perfectly edible. Ensure good air circulation by spacing plants at the recommended distance and opening vents in the greenhouse to reduce humidity and deter fungal infection.

BLOSSOM END ROT

PROBLEM Brown patches on the bases of tomato, pepper, and aubergine fruits.
CAUSE Insufficient calcium reaching fruits, caused by a lack of water.
REMEDY Affected fruits can't be saved, but watering consistently and never allowing the soil to dry out will keep fruits healthy. Water plants in pots at least once a day in summer.

CUCUMBER MOSAIC VIRUS

PROBLEM Foliage of cucumbers, squashes, and courgettes curls and develops yellow markings. Growth may be stunted.
CAUSE A common plant virus, often spread by aphids.
REMEDY Destroy affected plants and keep the garden free of weeds. Keep aphid numbers under control. Grow resistant varieties when available.

FRUIT SPLITTING

PROBLEM The skin of ripe tomatoes splits open or cracks.
CAUSE Fluctuations in temperature or the supply of water.
REMEDY Water regularly to maintain consistent moisture levels in compost and close greenhouse vents at night to help control temperature. Pick fruit as soon as it is ripe.

Freshly pulled carrots are crisp and bursting with sweet flavour. They are often an unexpected highlight of the summer harvest for new gardeners.

ROOT VEGETABLES

Fresh roots may not be the most glamorous crops, but they offer up big flavours – sweet, earthy, peppery, and always delicious. They are staples in hearty soups and stews but many also work as colourful ingredients in summer salads.

YEAR-ROUND HARVEST

Chitting seed potatoes signals the start of a new growing season in late winter. Speedy radishes are ready to pluck in spring and are followed by a succession of baby beetroot, turnips, kohlrabi, and carrots quickly in late spring and early summer, when the first waxy new potatoes can also be unearthed. Mature root veg are ready for harvest in summer and autumn, and most store well in a cool, frost-free place. Winter is the season for parsnips; their long, creamy roots stand in the soil until needed, sweetened rather than spoiled by frosts.

HOW MUCH SPACE?

Root crops need room in the soil to flourish. This can be a challenge in small gardens, but a few space-saving hacks make it possible to fit them in. They all, for example, grow well in deep pots, but you'll need a big container for potatoes and long-rooted parsnips. Radishes, turnips, and kohlrabi are made for intercropping (*see p.23*), and can yield a quick crop between slow-growing vegetables in as little as six weeks. Successional sowings of beetroot and carrots produce delicious baby roots that can also be harvested quickly. Early potato varieties do take up significant space, but once they are lifted early in summer, there is still time to plant follow-on crops for autumn and winter in their place.

WHAT THEY NEED

Success with root veg is all about providing the right soil conditions. Good drainage is essential, because waterlogged soil will impair growth and cause rot. Root crops benefit from a steady supply of moisture throughout growth and will only reach a satisfying size in fertile ground. Adding organic matter to the soil every year – ideally in the autumn – improves drainage, moisture retention, and fertility, which noticeably increases yields of roots season on season. Plants growing in good soil will only need watering during hot summer conditions or at specific stages of growth to help roots swell, but take care not to overwater as this can cause roots to split.

CARROTS

Newly pulled carrots have a distinctive aromatic, earthy flavour quite unlike roots available in the shops. The sheer number of varieties – and the diversity of shapes and colours – makes growing carrots great fun, but you'll need to be constantly vigilant for carrot fly, a common pest.

DIFFICULTY Easy to moderate
WHEN TO SOW Mar to July
IDEAL SOIL TYPE Light, deep, well-drained, and stone-free
SITE REQUIREMENTS Open, in full sun
GERMINATION TIME 10–21 days
GROW FROM Seeds
YIELD 3–5kg (7–11lbs) from a 2m (6ft) row

CALENDAR

	WINTER			SPRING			SUMMER			AUTUMN		
SOW				■	■	■	■	■				
HARVEST							■	■	■	■	■	

Time between sowing and harvesting
12–18 weeks

Grow bags filled with compost can be used for carrots if your soil is unsuitable. The bags should be at least 45cm (1½ft) deep.

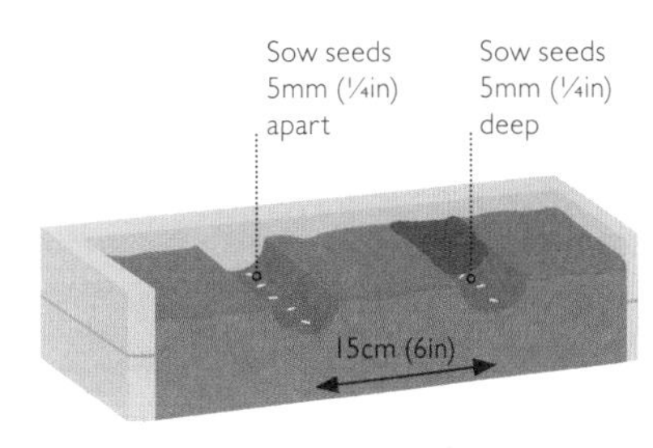

SOWING

Carrot seeds should be sown only when the soil has been warmed by the spring sun. To germinate reliably, they need light, sandy soil, well-raked to break down any lumps; if your garden has clay soil, you'll be much better off growing carrots in raised beds or containers.

Create a shallow trench about 5mm (¼in) deep, by laying down a cane and pushing it into the soil. Space rows 15cm (6in) apart and sow the seed thinly along the row; this avoids having to thin out seedlings later, which may attract carrot flies. Cover the seeds with soil; water through a rose.

Successive sowings in spring and summer will provide a continuous crop of roots through summer and autumn. Choose early varieties for quick crops of small roots and maincrops for larger carrots in late summer and autumn.

GROW

WEED Wispy carrot seedlings do not cope well with competition, so weed meticulously to keep the soil around them weed-free.

THIN Seedlings should be thinned in stages to about 5cm (2in) apart as they grow. Pinch them off at soil level and

Thin out carrot seedlings to allow the roots to grow to a good size.

ensure that all thinned seedlings are quickly disposed of to avoid attracting carrot flies, which are drawn to the scent of bruised carrot leaves.

PROTECT Slugs can be deterred using barriers or traps around young carrots (*see p.38*), but carrot flies are a more difficult problem. Unless you are lucky, you'll need to protect your crop against these tiny tunnelling maggots. No single method is foolproof:

- Try carrot varieties bred with some resistance, but don't rely on this alone.
- Cover rows of carrots completely with fine insect mesh or horticultural fleece to keep out the egg-laying adult flies.
- Adult flies are only capable of flying close to the ground. Erect a vertical barrier of polythene or fine insect mesh, at least 60cm (2ft) tall around the carrot bed.
- Keep carrots out of reach of adult flies by growing them in pots placed 60cm (2ft) above ground level.
- Try growing carrots alongside alliums such as onions, chives, or leeks; flies may be deterred by their scent.
- Don't grow carrots in the same place every year, because flies overwinter in the soil and may strike again in spring

WATER Carrots growing in the ground need minimal watering. Water those in pots consistently to keep the roots growing steadily and prevent splitting (which often occurs when carrots are overwatered after a dry spell).

Fence netting will prevent low-flying flies reaching the crop to lay their eggs.

HARVEST

It is difficult to judge when carrots are ready for harvest, because large roots may be fully concealed under the soil. Baby carrots are typically ready about two months after sowing early varieties. Check by feeling under the soil surface for the top of the root. Unearth carrots that have a reasonable width with a firm pull on the base of the leaves. You may need a fork to lift larger roots. Roots keep well in the ground but should be lifted before the first hard frost.

After harvest, you can store maincrop carrots by twisting off their foliage and placing them in boxes of damp sand in a cool, dark place.

VARIETIES

Early carrot varieties mature quickly and generally have fairly small roots that cannot be stored. Maincrop varieties take a few weeks longer to mature, but generally produce larger roots that can be stored successfully.

'AMSTERDAM FORCING' A tried and tested early carrot that matures rapidly to form long roots with a sweet flavour.

'ATLAS' Short, globe-shaped roots that are resistant to splitting make this early variety ideal for growing in containers.

'FLYAWAY' This full-flavoured early variety has some resistance to carrot fly.

'NORWICH' Reliable and vigorous, this is a maincrop Nantes variety with delicious, cylindrical roots up to 20cm (8in) long.

'RAINBOW MIX' A colourful combination of white, yellow, and purple maincrop roots means that you never quite know what you will unearth next.

BEETROOT

These sweet, earthy roots grow well in the soil or deep containers, where their red-stemmed leaves look pretty among summer flowers. A range of varieties gives you a choice of red, yellow, or patterned roots that add striking colour to salads, roasted veg, and pickles.

DIFFICULTY Easy
WHEN TO SOW Mar to July
IDEAL SOIL TYPE Fertile and well-drained
SITE REQUIREMENTS Full sun or light shade
GERMINATION TIME 10–14 days
GROW FROM Seeds
YIELD 2.5kg (5½lb) per 2m (6ft) row

CALENDAR

	WINTER			SPRING			SUMMER			AUTUMN		
SOW				■	■	■	■	■				
HARVEST							■	■	■	■	■	

Time between sowing and harvesting
9–12 weeks

SOWING

Beetroot can be sown successionally outdoors from early spring to midsummer. Cold conditions can cause early sowings to bolt. Prevent this by sowing seed in modules indoors, using bolt-resistant varieties, or covering young plants with fleece or cloches.

Draw out a row 2cm (1in) deep and sow seeds at 2cm (1in) intervals along it. Cover with soil, label, and water well. Space rows 20cm (8in) apart. Sow at the same depth in large, deep containers, spacing seeds 5cm (2in) apart. Alternatively, sow single seeds in modules and germinate on a cool windowsill or in a greenhouse.

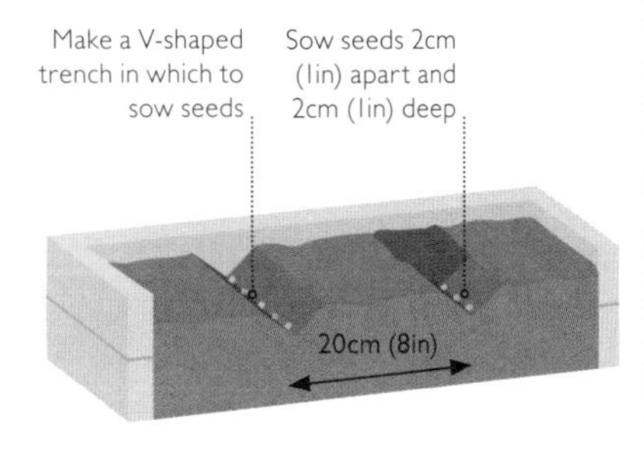

Water beetroot at all stages of development to keep the soil moist and encourage roots to swell.

GROW

Beetroot has multi-germ seeds, each of which will produce several seedlings. Pinch off weaker seedlings to thin to 7.5–10cm (3–4in) apart. Harden off module-raised seedlings and plant out at the same spacing in rows 20cm (8in) apart. Weed rows regularly and water to prevent the soil drying out. Create traps and barriers to protect seedlings from slugs and snails.

To harvest beetroot, grip the stems and pull the roots up from the soil.

HARVEST

Beetroots can be picked young, as soon as they reach a usable size, or left to grow to the size of a tennis ball. The leaves and stems are also delicious. Harvesting alternate plants encourages remaining roots to swell. Large roots harvested in mid-autumn can be stored in sand-filled boxes in a frost-free shed.

VARIETIES

Jazz up your salads with the bold colours of several beetroot varieties. Choose a bolt-resistant variety for early sowings.

'BOLTARDY' A dark red, bolt-resistant variety sown through spring and summer.

'CHIOGGIA' Dramatic pink and white-striped roots look spectacular in salads.

'PABLO' Produces uniform, spherical roots and attractive upright foliage.

RADISHES

The ultimate fast food, radishes can be ready to pick just four weeks after sowing, which makes them handy to squeeze in between slower-growing crops to maximize space. They are incredibly easy to grow, but keep the soil moist, or the crisp, peppery roots develop a fiery heat.

DIFFICULTY Easy
WHEN TO SOW Mar to Aug
IDEAL SOIL TYPE Moist and well-drained
SITE REQUIREMENTS Full sun or light shade
GERMINATION TIME 3–7 days
GROW FROM Seeds
YIELD 40 radishes per 1m (3ft) row

CALENDAR

	WINTER	SPRING	SUMMER	AUTUMN
SOW				
HARVEST				

Time between sowing and harvesting
4–6 weeks

SOWING

Fast-growing radishes tend to mature all at the same time, so sow small amounts every 2–3 weeks from early spring to late summer for a continuous crop. Rake your soil to remove any stones and lumps and create a shallow trench 1cm (½in) deep with a cane. Sow seed thinly at 2.5cm (1in) intervals along the row to reduce the need for thinning later; cover them with soil and water well. Space rows 15cm (6in) apart.

Make summer sowings in light shade to help stop plants running to seed (bolting). Radishes can also be sown in large pots, with seeds spaced at 2.5cm (1in) intervals.

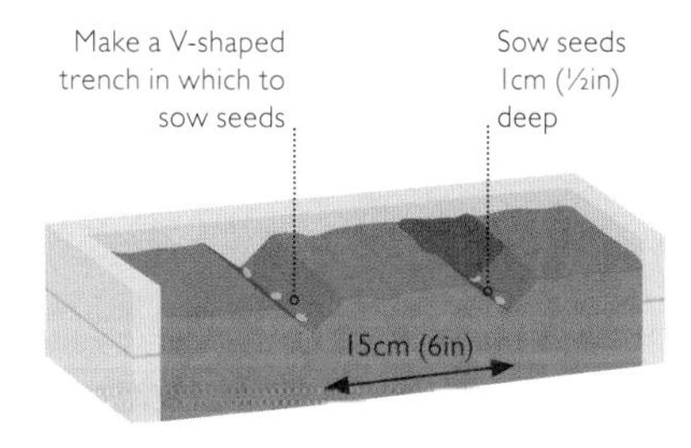

Thinning radish seedlings helps to ensure that the radishes you harvest are plump and juicy.

GROW

Seedlings grow very rapidly and need to be thinned to 2.5cm (1in) apart promptly to avoid checking their growth. Water radishes in the soil thoroughly once a week during dry weather and those in pots daily to prevent them bolting and keep roots succulent. Grow under fleece to prevent problems with flea beetle.

HARVEST

Pick radishes by pulling them from the soil by their leaves as soon as they are the size of large marbles. Harvest quickly because plants tend to bolt rapidly and mature roots become woody. Radishes make a satisfying snack or a colourful addition to salads.

Traditional varieties have red cylindrical roots with a peppery flavour.

VARIETIES

Varieties are long or spherical, ranging in colour from red and purple to white.

'AMETHYST' Beneath unusual purple skin, these spherical roots have white flesh.
'CHERRY BELLE' This reliable radish yields red, globe-shaped roots.
'SPARKLER' An easy variety producing spherical roots with attractive white tips.

TURNIPS

Turnips are a versatile, fast-maturing crop, easy to grow and well-suited to intercropping among veg that take longer to develop. Successional sowings of early and maincrop varieties can be harvested almost year-round when grown for their tasty green tops as well as their roots.

DIFFICULTY Easy
WHEN TO SOW Apr to July (early varieties); July (maincrop varieties); Aug to Sept (turnip tops)
IDEAL SOIL TYPE Fertile, well-drained, and moist
SITE REQUIREMENTS Full sun or light shade
GERMINATION TIME 4–7 days
GROW FROM Seeds
YIELD 2.5kg (5½lb) per 2m (6ft) row

CALENDAR

	WINTER	SPRING	SUMMER	AUTUMN
SOW				
HARVEST				

Early varieties
Maincrop varieties
For turnip tops

Time between sowing and harvesting
6–12 weeks

Turnip leaves are rich in vitamins and taste similar to mustard greens.

SOWING

Turnips are best sown directly into open soil, but can also be grown in large containers. Wait until the soil is warm in mid-spring, because turnips tend to run to seed (bolt) when sown too soon.

Sow early varieties successionally every 3–4 weeks from mid-spring to midsummer, in rows 1cm (½in) deep and 23cm (9in) apart. Slower-maturing maincrop varieties can be sown at the same depth in midsummer, in rows 30cm (12in) apart. For a spring crop of turnip tops – greens with a distinctive peppery, lemony flavour – sow a maincrop variety in late summer or early autumn in rows 15cm (6in) apart.

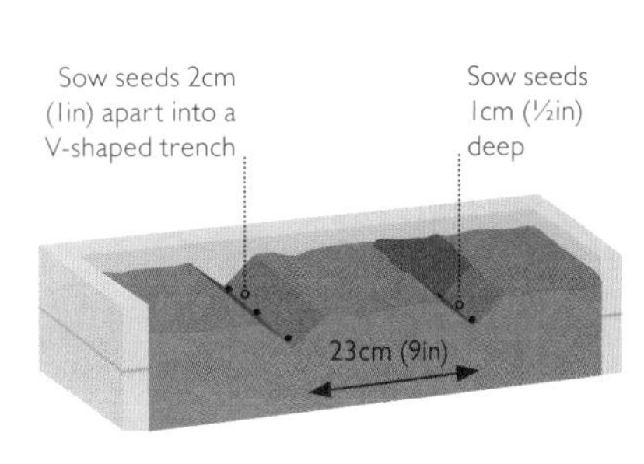

Water consistently to prevent turnip roots from becoming bitter and tough.

GROW

Thin the fast-growing turnip seedlings promptly to 10cm (4in) apart for early varieties and 15cm (6in) intervals for maincrops; this will give their roots the space they need to fully develop. Weed around plants regularly and keep the soil moist by watering thoroughly once a week in summer. Protect young plants with fleece before flea beetles damage foliage.

HARVEST

Pick early varieties as soon as six weeks after sowing, when they reach the size of a golf ball. Maincrop varieties will stand for longer without getting tough, and are hardy enough to be left in the soil during winter. Cut leafy turnip tops when they are about 15cm (6in) tall.

VARIETIES

Choose early varieties for quick summer crops and maincrop varieties for winter roots and spring tops.

'GOLDEN BALL' A hardy maincrop variety with spherical, yellow-fleshed roots.
'PURPLE TOP MILAN' This fast-growing plant has flattened roots best eaten small.
'SNOWBALL' Good crops of elegant white, globe-shaped roots can be harvested from this early variety.

PARSNIPS

Parsnips need a long growing season, but it is definitely worth the wait for a supply of aromatic, freshly lifted roots through autumn and winter. They are undemanding during growth and happy to stand in the soil once mature, which makes them an ideal crop for beginners.

DIFFICULTY Easy
WHEN TO SOW Mar to May
IDEAL SOIL TYPE Deep and well-drained
SITE REQUIREMENTS Full sun
GERMINATION TIME 14–21 days
GROW FROM Seeds
YIELD 4kg (9lb) per 2m (6ft) row

CALENDAR

	WINTER	SPRING	SUMMER	AUTUMN
SOW				
HARVEST				

Time between sowing and harvesting
5 months

SOWING

To ensure good germination, use fresh seed each year, sowing it directly into the ground after the soil has warmed in spring; in colder areas, warm up the soil before sowing by covering it with fleece or cloches.

Rake the soil to a fine crumb and create a trench 1cm (½in) deep across the bed. Sow clusters of three seeds at 15cm (6in) intervals along the row. Space rows 30cm (12in) apart. Cover the trench with soil, label, and water thoroughly. To make the best use of space, sow fast-growing radishes thinly between the seed clusters.

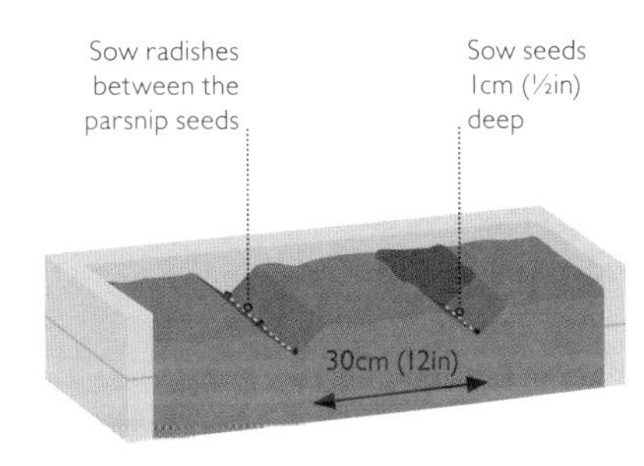

Pull up weeds by hand to avoid damaging the parsnip roots.

GROW

Thin seedlings to one per station by pinching off or uprooting the weaker plants with your fingers. Cover with fleece where there is a risk of carrot fly damage. Weed regularly, particularly while plants are young, being careful not to damage their developing roots. Harvest the radishes as soon as they are ready, allowing the parsnips room to grow. Watering is needed only during long spells of hot, dry weather.

HARVEST

Parsnip roots are ready to lift carefully with a fork as soon as the leaves start to die down; however, they are commonly left longer because exposure to frost improves their flavour. Mark the row so that it is easy to find without the foliage.

Parsnips can be left in the ground and harvested as required through winter.

VARIETIES

Parsnips all have a similar flavour, but differing shapes. Opt for those bred with resistance to parsnip canker.

'GLADIATOR' A vigorous, smooth-skinned parsnip with a sweet flavour.
'JAVELIN' High yielding and reliable, this long variety stands in the soil well.
'TENDER AND TRUE' Long, straight roots are rarely troubled by canker.

POTATOES

The sweet, waxy flesh of home-grown new potatoes betters anything on supermarket shelves, and unearthing these tiny gems in early summer is a highlight of the growing season. They are so easy to grow from "seed potatoes" that you will wonder why everyone doesn't do it.

DIFFICULTY Easy
WHEN TO SOW Mar to Apr (early varieties): Apr to May (maincrop varieties)
IDEAL SOIL TYPE Fertile and well-drained
SITE REQUIREMENTS Full sun
GERMINATION TIME 14–21 days
GROW FROM Seed potatoes
YIELD 3–5kg (6½–11lb) per 2m (6ft) row

CALENDAR

	WINTER	SPRING	SUMMER	AUTUMN
SOW				
HARVEST				

Early varieties
Maincrop varieties

Time between planting and harvesting
3–5 months

Potatoes will start to develop shoots after about two weeks of chitting.

BEFORE PLANTING

Potatoes are grown from specially raised, certified disease-free tubers called "seed potatoes", which are on sale in garden centres or by mail order during mid- to late winter. A 1kg (2lb) bag of seed potatoes should fill two 2m (6ft) rows. About four weeks before planting, you should start "chitting" the seed potatoes – a process that ends the dormancy of the buds (commonly called the "eyes" of the potato) and triggers the growth of shoots. To do this, stand the potatoes in an egg box or seed tray so that the end of the tuber with the most eyes faces upwards. Place the tray in a cool, bright room and leave the tubers to form sturdy, dark sprouts. Chitting is important to start early varieties into growth, but is less critical for maincrop varieties.

EARLY OR MAINCROP?

Early potato varieties are quicker to crop and more compact than maincrop varieties but produce lower yields, so suit smaller plots better. They are commonly subdivided into "first earlies" (which are the quickest growing of all) and "second earlies". Maincrop potatoes produce larger harvests that will store well into winter, but those growing in late summer are more prone to potato blight – a damaging fungal disease.

POTATO CROP PLANTING GUIDE

	EARLY	MAINCROP
Time to harvest	75–110 days	135–160 days
Planting distance	30cm (12in)	38cm (15in)
Distance between rows	50cm (20in)	75cm (30in)
Yield	3kg (6½lb) from 2m (6ft) row	5kg (11lb) from 2m (6ft) row

PLANTING

Potato plants are frost tender, so it is particularly important to pay attention to the weather and wait for the soil to warm in spring before planting. If possible, cover the soil with fleece for two weeks before planting in cooler areas; it is far preferable to keep tubers chitting for another week or two than have them struggle in cold, wet ground. Choose a sunny position, where plenty of organic matter was added the previous autumn. Plant earlies from early to mid-spring and maincrops from mid- to late spring.

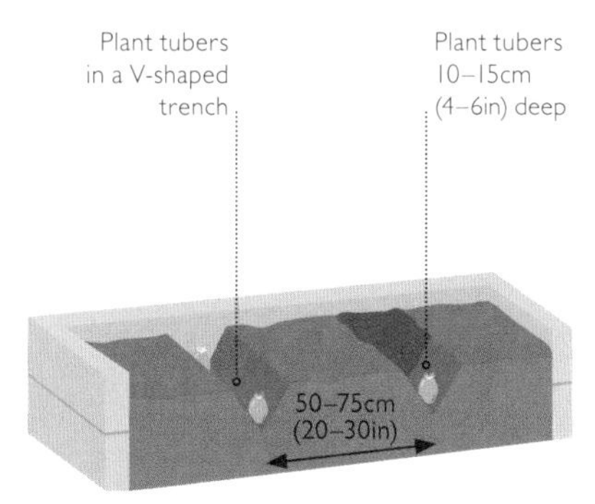

IN THE SOIL Make a V-shaped trench 10–15cm (4–6in) deep across the bed with a spade or rake. Alternatively dig a hole to that depth for each individual tuber along the row. Space early varieties 30cm (12in) apart in rows 50cm (20in) apart, and maincrops at 38cm (15in) intervals along rows 75cm (30in) apart. Place tubers into the trench or hole with their shoots pointing upwards and cover them with soil, taking care not to damage them.

IN CONTAINERS Early varieties will thrive in containers least 30cm (18in) wide and deep. They will also do well in sturdy bags (often sold as potato planters); indeed, this is a good way to grow an extra-early crop in a greenhouse. Ensure that the containers have drainage holes. Add a 10cm (4in) layer of multi-purpose compost to the base of the pot, place one sprouted tuber on the surface (or more in a larger container) and cover with a 10cm (4in) layer of compost.

Before planting rub off all but 3–4 of the strongest shoots or you'll get lots of small potatoes.

A container the size of a dustbin is sufficient for about 4–5 seed potatoes.

GROW

PROTECT Emerging foliage is easily damaged by frost, so cover shoots with fleece, cloches, or even newspaper if a cold night is forecast.

EARTH UP Once plants are 15–20cm (6–8in) tall, earth them up by gently pulling soil around their stems with a spade or rake. Leave just the top leaves showing above this soil. This shields the plant from frost and helps prevent developing tubers from being revealed at the surface where they turn green and poisonous on exposure to light. A generous covering of soil also affords protection from infection with potato blight. Earth up plants in stages until the ridge is around 30cm (12in) tall. Plants growing in containers should also be earthed up by progressively adding 10cm (4in) layers of compost as they grow, until the compost is just below the rim of the container.

WATER Potato plants need a plentiful supply of water to produce a good yield, but those growing in the ground need additional watering only during dry spells of weather. Once the foliage is fully grown and tubers have started to form, however, generous watering is important for a good harvest. Container-grown plants may require daily watering in hot weather, but check moisture levels in the compost because overwatering can cause tubers to rot.

To earth up, pull soil from two sides of the potato plant to form a ridge.

HARVEST

Early varieties should be ready to harvest in early to midsummer, once their flowers are open. Push some soil aside with your hand to assess the tubers and lift them carefully with a fork as required once they reach a usable size. Maincrops can be harvested fresh in the same way from late summer, but to produce tubers for storage, allow plants to continue growing into early autumn. Watch out for the symptoms of potato blight (see *p.125*) in late summer and immediately remove diseased foliage to prevent the blight spreading to tubers. Cut down the tops of healthy plants when they yellow and leave the potatoes in the soil for ten days before unearthing them. Allow them to dry on the soil for a few hours before storing in thick paper sacks in a dark, cool, and frost-free spot.

Dry maincrop potatoes by spreading them out on the ground before storing.

WHY NOT TRY?

For a late crop of new potatoes to enjoy at the end of autumn, try "second crop" potatoes. These are ordinary seed potatoes that have been kept dormant in cold storage until midsummer, when they can be planted in the soil or large containers with no chitting, and grown in the same way as early varieties. Growing late in the year leaves them vulnerable to potato blight and the first frosts in colder regions, so choose varieties with some blight resistance, and protect them from the cold with fleece or by moving containers under cover. Potatoes will be ready to harvest in October, but those in pots can also be kept in a cool, frost-free shed or garage to unearth for fresh new potatoes at Christmas.

New potatoes can be grown right through to early winter in containers.

VARIETIES

Are you looking for a waxy salad potato or a floury variety to bake or roast? Consider this, along with timing and disease resistance, when choosing.

'ARRAN PILOT' A traditional first early, waxy, salad potato with firm white flesh, rich flavour, and resistance to scab.

'CHARLOTTE' This second early has firm, pale yellow flesh and resistance to scab.

'NICOLA' Bumper crops of small salad potatoes with yellow, waxy flesh make this second early a popular variety.

'PINK FIR APPLE' The pink-skinned, knobbly tubers of this maincrop variety have a nutty flavour and are delicious boiled or roasted.

'SARPO MIRA' This maincrop has good blight resistance, ensuring a healthy crop of pink-skinned, floury potatoes for storing.

'SWIFT' One of the earliest varieties to crop, it produces waxy, white-skinned salad potatoes. It is ideal for containers.

KOHLRABI

Weird and wonderful kohlrabi produce crisp swollen stems that have a delicate turnip-like flavour and are delicious grated raw into salads or gently steamed. Quick to crop and tolerant of a wide range of conditions, kohlrabi is an ideal plant for new gardens, containers, and intercropping.

DIFFICULTY Easy
WHEN TO SOW Feb (indoors), Mar to Aug (outdoors)
IDEAL SOIL TYPE Fertile, light, and well-drained
SITE REQUIREMENTS Full sun
GERMINATION TIME 5–10 days
GROW FROM Seeds
YIELD 18 bulbs per 2m (6ft) row

CALENDAR

	WINTER			SPRING			SUMMER			AUTUMN		
SOW			Indoors	Outdoors	Outdoors	Outdoors	Outdoors	Outdoors	Outdoors			
HARVEST						Indoors/Outdoors	Indoors/Outdoors	Indoors/Outdoors	Indoors/Outdoors	Indoors/Outdoors	Indoors/Outdoors	Indoors/Outdoors

Indoors
Outdoors

Time between sowing and harvesting
9–12 weeks

SOWING

Sow into warm soil in spring, because kohlrabi tend to run to seed (bolt) if sown when it is cold. For an early start in late winter or early spring, sow in modules indoors to transplant outside once the weather becomes milder.

Sow single seeds 1cm (½in) deep into modules filled with multi-purpose compost. Water, label, and place on a warm windowsill to germinate. Sow outdoors from early spring until late summer by sprinkling seed thinly along a trench 1cm (½in) deep. Cover with soil, water well, and label the row. Sow small quantities in succession every three weeks, in rows 30cm (12in) apart, for a continuous crop.

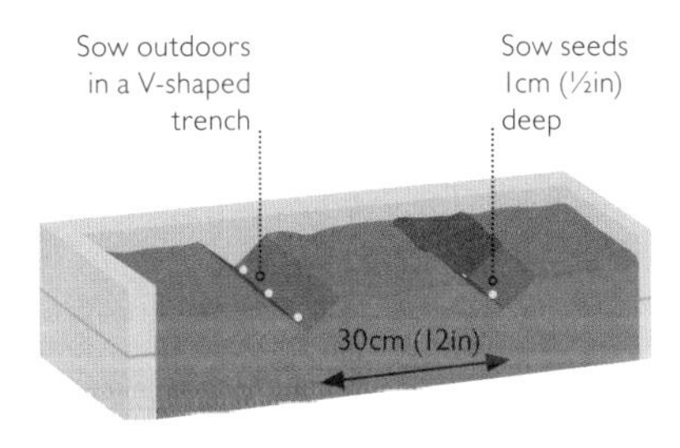

Harvest by cutting the bulb from the root at ground level with a sharp knife.

GROW

Harden off seedlings raised indoors once they are 5cm (2in) tall and plant them out at 10cm (4in) intervals in rows 30cm (12in) apart. Thin seedlings sown into the soil to 10cm (4in) apart while they are small. Keep the soil free of weeds and water regularly during dry weather. Protect plants with fleece as the weather cools in autumn.

HARVEST

Start pulling kohlrabi from the soil when they reach the size of a ping-pong ball. Small bulbs make the best eating; they are crisp and tender, but will quickly toughen if left to grow larger than a tennis ball. The leaves are also delicious as cooked greens.

Most kohlrabi varieties have green bulbs; purple bulbs are less common.

VARIETIES

Traditionally, green varieties are sown in spring for summer crops and purple varieties in summer for autumn harvest.

'KOLIBRI' Purple-skinned bulbs are white inside and produce quick crops through summer and autumn.

'KOSSAK' This sturdy, pale green variety forms large, tender bulbs.

TROUBLESHOOTING

You can avoid many root crop problems by delaying sowing until the soil is warm in spring, and by improving the soil with organic matter so that it is well-drained and moisture-retentive. Carrot fly and potato blight can ruin harvests, so speak to local gardeners about their experiences and use preventative measures wherever possible.

BEET LEAF MINER

PROBLEM Pale brown tunnels inside beetroot leaves that can set back growth on young plants.
CAUSE Fly larvae feeding on leaf tissue from inside the leaf.
REMEDY Remove affected leaves. Rotate beetroot around the garden and grow under insect mesh or fleece to prevent flies laying eggs on plants.

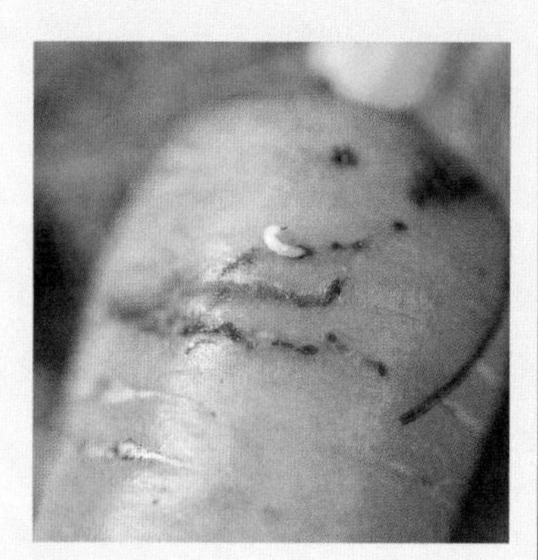

CARROT FLY

PROBLEM Thin brown tunnels ring the roots of carrots, often making them inedible. Parsnips are also affected.
CAUSE The small, white larvae of carrot flies feeding on roots.
REMEDY Avoid thinning, as the scent attracts flies to lay eggs. Exclude flies with insect mesh, fleece coverings, or barriers. Grow resistant varieties.

FLEA BEETLE

PROBLEM Tiny rounded holes in the foliage of radishes, turnips, and kohlrabi.
CAUSE Small beetles, which jump from leaves when disturbed.
REMEDY Cover vulnerable crops with fine insect mesh or fleece after sowing to keep the beetles out. Sow into warm soil and keep seedlings well watered to encourage rapid growth.

SLUGS AND SNAILS

PROBLEM Seedlings are destroyed and ragged holes appear in leaves. Roots are pitted or tunnelled.
CAUSE Feeding by slugs and snails, which are usually more active in wet weather.
REMEDY Set traps and create barriers around seedlings. Manually remove slugs and snails after dark. Harvest potatoes promptly.

BOLTING

PROBLEM Plants flower prematurely, preventing the formation of large roots.
CAUSE Bolting is triggered by sowing when conditions are too cold, or dry soil during growth.
REMEDY Avoid sowing too early in spring. Make early sowings in modules indoors. Grow bolt-resistant varieties. Water to keep soil moist.

PARSNIP CANKER

PROBLEM Orange-brown or purple patches of rot, usually starting near the top of parsnip roots.
CAUSE A fungal disease that enters the root through cracks or pest damage.
REMEDY Protect parsnip crops from carrot fly damage. Grow resistant varieties. Water to avoid soil drying out during hot summer weather.

POTATO BLACKLEG

PROBLEM Stem bases rot and turn black, causing stunted growth, yellowing foliage, and rotten tubers.
CAUSE A bacterial infection, which is worse in wet summers.
REMEDY Dispose of infected plants quickly. Ensure soil is well-drained, rotate potato crops around the garden each year, and grow resistant varieties.

POTATO BLIGHT

PROBLEM Brown patches of wet rot on leaves and stems spread until plants collapse. Tubers may develop rusty red marks, which soften and rot.
CAUSE A fungus-like organism.
REMEDY Earth up to prevent infection of tubers. Grow blight-resistant or early varieties that are harvested before blight spreads. Dispose of infected foliage.

POTATO SCAB

PROBLEM Raised, rough, brown marks on the skin of tubers, which can be removed before eating by peeling.
CAUSE Invasion of the potato skin by bacteria-like soil microorganisms.
REMEDY Scab is worse in dry conditions, so improve moisture retention in soil by adding organic matter, and water crops regularly. Grow resistant varieties.

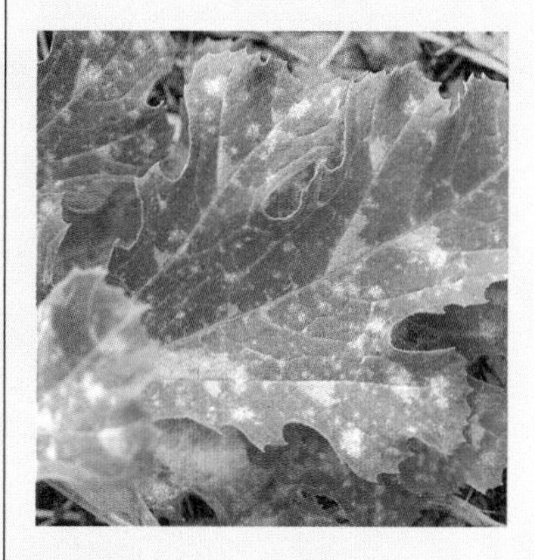

POWDERY MILDEW

PROBLEM A dusty, white covering appears on the surfaces of leaves; turnips are particularly susceptible.
CAUSE Infection by fungal spores.
REMEDY Thin to the recommended spacing to allow air to circulate. Plants stressed by lack of water are more prone, so water regularly during dry weather, especially on sandy soils.

Glorious red stems make ruby chard a star of the veg plot, but all leafy greens have ornamental value.

LEAFY VEG

These are the tough crops at the heart of any veg plot. They thrive in cool, moist conditions and are often hardy enough to stand through winter. Grow a range of leafy veg for autumn stir-fries, rich winter sides, and tender spring treats.

A FEAST OF GREENS

Sow some fast-growing leafy crops alongside brassicas for a variety of veg to harvest throughout the year. The sprinters among greens are iron-rich spinach, earthy chard, and peppery oriental leaves, and several sowings will guarantee generous pickings through summer and autumn. Spring cabbage, kale, and sprouting broccoli, on the other hand, have marathon growing seasons, but make up for this in hardiness; kale can be picked through the dark days of winter and sprouting broccoli and spring cabbage provide a tender harvest early in the year, when new spring sowings are only just going into the soil.

HOW MUCH SPACE?

Leafy greens look spectacular in the ordered rows of a vegetable plot, but their vibrant colours and varied textures and forms earn them a place in container displays and ornamental borders too. Spinach, chard, and oriental greens grow just as successfully in well-watered containers as they do in open soil, and are highly productive, especially when outer leaves are picked as needed, leaving the plants to continue growing. Sprouting broccoli, cabbages, and kale only really flourish in open soil, but you need very few of these plants for a good crop; with a little planning they will follow on perfectly after early vegetables, such as broad beans or peas, making it easier to find space for them, even in small gardens.

KEEPING CROPS HEALTHY

Leaf crops thrive in fertile soil that is rich in moisture-retaining organic matter – vital to prevent bolting in hot, dry conditions. A garden brimming with tender foliage attracts pests, but plenty can be done to reduce pest damage. Create traps and barriers to catch slugs and snails tempted by young plants and collect the critters by torchlight to control their numbers. Members of the cabbage family may also need protection from flea beetle and cabbage root fly while young, and a covering of netting as they mature will keep away egg-laying butterflies and hungry pigeons.

SPINACH

Spinach is easy to grow, provided that you sow it at the right time of year. It thrives if sown during spring or autumn, but if sown in summer will quickly run to seed. Its nutritious, deep green leaves can be picked when small for salads or left to mature for cooking.

DIFFICULTY Easy
WHEN TO SOW Mar to May (summer varieties); Aug to Sept (winter varieties)
IDEAL SOIL TYPE Fertile, moisture-retentive
SITE REQUIREMENTS Open, sunny
GERMINATION TIME 5–21 days
GROW FROM Seeds
YIELD Up to 1kg (2lb) from a 2m (6ft) row

CALENDAR

	WINTER			SPRING			SUMMER			AUTUMN		
SOW												
HARVEST												

Summer varieties
Winter varieties

Time between sowing and harvesting
6–10 weeks

SOWING

Avoid sowing spinach in summer: sow summer varieties from early spring and winter varieties in autumn. In either case, dig in plenty of well-rotted compost before sowing: this will help to loosen the soil (spinach has deep roots and fares poorly in compacted soils) and retain the moisture that spinach loves. Draw out a trench 2.5cm (1in) deep using a cane and sow seeds at intervals of roughly 1cm (½in). Label the row, cover over with soil, and water thoroughly using a watering can fitted with a rose. Space rows 30cm (12in) apart if you want mature leaves; for baby spinach, space the rows 15cm (6in) apart. Stagger your harvest by sowing at intervals of one or two weeks throughout the spring or autumn. Seeds can be sown into containers, as long as they are large enough to retain moisture. Container-grown spinach benefits from some protection from slugs and snails.

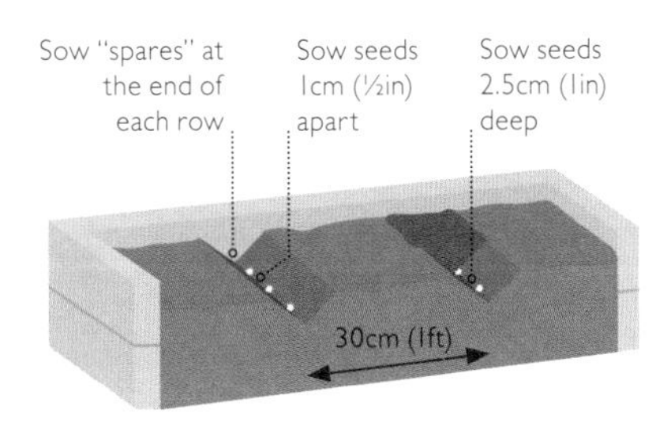

Grow spinach as a salad crop in a deep container and harvest the leaves regularly.

GROW

WATER Spinach likes a lot of water, but not saturated soils. Make sure never to let the soil dry out completely. Water around the plants regularly, particularly as the weather warms in late spring. Keeping the soil moist produces strong growth and helps to prevent plants flowering prematurely (bolting), which stops the production of new leaves and makes those that remain taste bitter. Spinach growing in containers will need watering frequently – possibly daily or even twice daily in summer – to keep the compost moist

Water seedlings little and often to keep the soil moist but not soggy.

THIN For large, mature spinach plants, gradually thin seedlings to about 15cm (6in) apart, by pinching them off at soil level. Make use of the thinnings in salads. There is no need to thin rows of seedlings grown for baby leaves.

PROTECT Slugs and snails adore spinach, so it is essential to create traps or barriers around your rows to keep them at bay (*see p.38*). Autumn-sown spinach can be left in the soil over winter to crop again in spring, but will need the protection of cloches, or a mulch of dry straw or leaves around the base of plants, when the weather turns cold.

Overwintering spinach plants can be mulched with dry autumn leaves.

HARVEST

Leaves can simply be picked from the outside of the plants as required, leaving those at the centre of the rosette to keep growing. Pick baby leaves when they are about 10cm (4in) high, to eat raw in salads. Larger, thicker leaves need to be picked in quantity for cooking, because they wilt down significantly in the pan.

Harvest spinach by pinching off the outer leaves and working your way in as leaves mature; alternatively, simply cut the plant at the base.

WHY NOT TRY?

Amaranth or callaloo makes a colourful and nutritious summer alternative to spinach. As the plant grows, you'll need to pinch out the central stem to promote bushy growth and remove any flowers that form. Be sure to choose a variety used for leaf production (rather than grains). This tender, heat-loving plant needs a sheltered spot outdoors to flourish and is best sown in summer, directly into soil that has warmed under fleece for at least a week.

Callaloo seeds are tiny, so mix them with sand to make them easier to handle. Sow the seeds 0.5cm (¼in) deep, spaced 10cm (4in) apart for baby leaves, or 30cm (12in) apart to produce mature leaves on plants up to 1.2m (4ft) tall. Thin the seedlings and harvest the crop as you would spinach leaves.

Some varieties of callaloo have decorative pink-tinged leaves.

VARIETIES

Sow spinach varieties at the recommended time of year to ensure success; choose summer varieties for spring sowings and winter varieties to sow in late summer and autumn.

'GIANT WINTER' A hardy winter spinach that yields large, slightly puckered leaves.

'KOOKABURRA' This upright variety has flavoursome dark oval leaves.

'MATADOR' Perfect for picking as baby leaves: make successional sowings of this summer variety throughout spring.

'MEDANIA' This variety is good for overwintering, producing smooth leaves in spring. It has good resistance to bolting so can be used for late spring sowings.

'TETONA' Slow to bolt, with delicious leaves, this variety is best for spring sowings.

CHARD

Chard, or leaf beet, is the easiest leafy crop to grow. It is usually untroubled by pests and diseases, tolerates drought, and can be harvested year-round from just two sowings. Its coloured stems and leaf veins, in shades of yellow, pink, red, and white, look glorious in the garden and on the plate.

DIFFICULTY Easy
WHEN TO SOW Mar to July
IDEAL SOIL TYPE Moist and fertile
SITE REQUIREMENTS Full sun or light shade
GERMINATION TIME 7–14 days
GROW FROM Seeds or plants
YIELD 2kg (4lb) per 2m (6ft) row

CALENDAR

	WINTER			SPRING			SUMMER			AUTUMN		
SOW				■	■	■	■	■				
HARVEST	■	■	■	■	■	■	■	■	■	■	■	■

Time between sowing and harvesting
8–12 weeks

SOWING

Sow chard at wide spacings to produce large leaves for cooking or much closer together for baby salad leaves. Don't sow too early, as plants sown in the cold tend to flower prematurely (bolt).

FOR MATURE LEAVES Make two sowings to supply large leaves throughout the year; one in mid-spring to crop in summer and autumn, and one in midsummer to pick from autumn to the following spring.

Stretch a string across the bed and make a trench 2cm (1in) deep along it. Sow single seeds at 10cm (4in) intervals, cover with soil, water well, and label. Space rows 38–45cm (15–18in) apart.

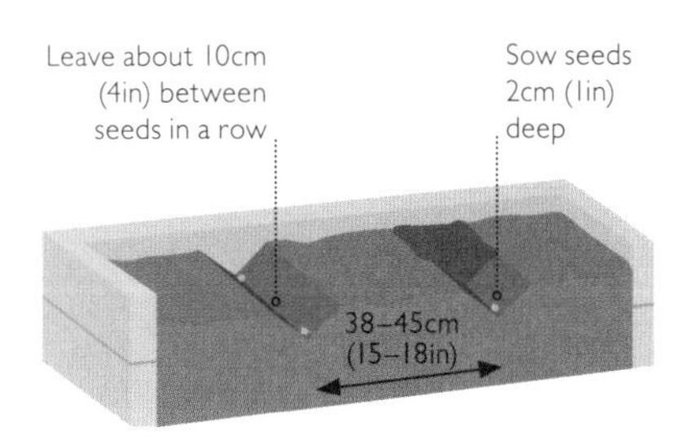

Sowing in modules will give your chard plants a head start for summer salads.

Plants can be raised in modules, on a windowsill, or outdoors, for transplanting when 5cm (2in) tall. Sow one seed 2cm (1in) deep in the centre of each module, then cover with compost. Water well.

FOR BABY SALAD LEAVES Between early spring and midsummer, sow successionally every 2–3 weeks into soil or large containers. Sow seeds 2cm (1in) deep, into single rows, spacing seeds roughly 5cm (2in) apart and rows 20cm (8in) apart; or sow into broad 15cm- (6in-) wide drills, spacing seeds roughly 5cm (2in) apart each way.

GROW

THIN Each chard "seed" is actually a cluster of seeds that will produce several seedlings. Gradually thin them to 30cm (12in) apart for mature plants or 5cm (2in) apart for baby leaves. Thin chard sown in modules to one plant per cell after planting out.

Transplant seedlings by holding the rootball or leaves rather than the stem.

PLANT OUT Harden off young chard plants raised indoors by gradually acclimatizing them to outdoor conditions over a period of 7–10 days. Plant them out at 30cm (12in) intervals in rows 38–45cm (15–18in) apart if you want to harvest mature leaves, or at 5cm (2in) intervals in rows 20cm (8in) apart if your goal is to pick baby leaves.

Water red-stemmed varieties often as they are most prone to bolting.

WATER Chard plants develop deep roots and can survive long dry spells without watering. However, they will yield more leaves given a thorough soak weekly during summer to keep the soil moist. Container-grown plants need regular watering. Apply a spring mulch around plants to retain moisture and suppress weed growth.

PROTECT Protect seedlings from slugs and snails with barriers and traps. Birds sometimes damage seedlings; cover with netting or fleece if this is a problem. For better-quality leaves to harvest during winter, cover rows with fleece tunnels.

HARVEST

Mature leaves are ready about 12 weeks after sowing. Start picking them from the outside of each plant by snapping or cutting the base of each stem. Salad leaves can be harvested in the same way after about eight weeks. Pick off any flower stems that form to prolong the harvest.

Harvest gradually as needed by cutting or snapping off the outer stems.

Harvest mature plants in one go by cutting the stems at the base.

WHY NOT TRY?

New Zealand spinach is a great leaf crop if your soil is too poor, or your site too sunny, for chard or spinach. Sow indoors in small pots in mid- to late spring to plant outdoors in early summer. Soak the seeds in water for 12 hours, then sow 1cm (½in) deep, cover with compost, water, and place on a windowsill. Harden off before planting out about 45cm (18in) apart in a sunny position with well-drained soil. The fleshy, triangular leaves at the shoot tips can be picked repeatedly through summer and autumn.

New Zealand spinach is a low-growing plant with a mild taste.

VARIETIES

Grow a range of bold colours to brighten up your vegetable plot.

'BRIGHT LIGHTS' A brilliant mix of ornamental red, pink, yellow, and white stems which is often grown for salad leaves.

'BRIGHT YELLOW' The rich gold of the thick stem continues into the leaf veins.

'FORDHOOK GIANT' A large, white-stemmed variety that grows up to 60cm (2ft) tall.

PERPETUAL SPINACH OR SPINACH BEET Grown as a hardier, more bolt-resistant alternative to true spinach, this vigorous variety has slim green stems and glossy green leaves.

SPROUTING BROCCOLI

A crop of delicious, tender broccoli shoots is the ideal way to fill the "hungry gap" between winter and the first harvest of spring-sown vegetables. These plants are large and prolific, so you may only need two or three. Grow "tenderstem" varieties to pick in summer and autumn.

DIFFICULTY Easy
WHEN TO SOW Apr to June
IDEAL SOIL TYPE Fertile, well-drained and firm
SITE REQUIREMENTS Full sun and sheltered
GERMINATION TIME 4–7 days
GROW FROM Seeds or plants
YIELD 1kg (2lb) per 2m (6ft) row

CALENDAR

	WINTER	SPRING	SUMMER	AUTUMN
SOW				
HARVEST				

Tenderstem
Hardy varieties

Time between sowing and harvesting
3–7 months

SOWING

Sprouting broccoli has a long growing season, so must be sown in a seedbed or raised in modules to transplant into its final position in summer. Sowing both early and late varieties will give you stems to cut over a long period. If you only need a few plants, buy them from a garden centre or by mail order.

SOWING INTO SOIL From mid-spring to early summer, raise a small number of plants in a seedbed. Create a trench 1cm (½in) deep and 60cm (2ft) long using a trowel; sow seed thinly along it, cover with soil, water, and label. Space rows 20cm (8in) apart. After germination, thin seedlings to 5cm (2in) apart.

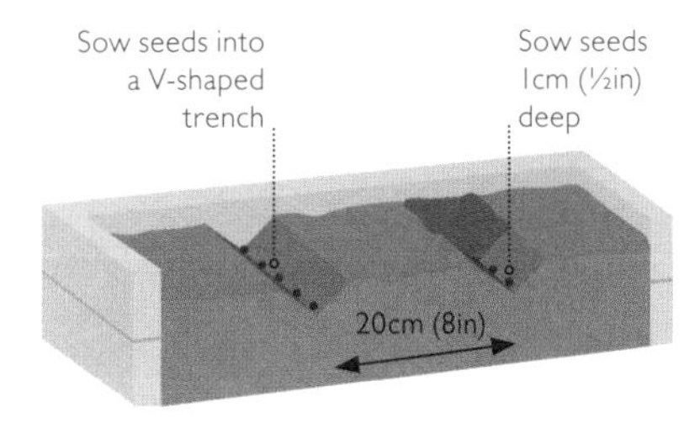

Sow two seeds in each module to increase your chances. Remove the weaker seedling should both germinate.

SOWING IN MODULES Fill a module tray with multi-purpose compost and sow seeds 1cm (½in) deep in the centre of each module. Cover with compost, water, and carefully label to distinguish between early and late varieties, if growing both. Place on a sunny windowsill in a cool room or in a sheltered place outdoors. Transfer the seedlings into 9cm (3½in) pots once they are large enough to handle safely.

GROW

TRANSPLANT Sprouting broccoli dislikes acid soils, so you may need to add lime (*see p.20*) the year before planting in the final position. Harden off indoor-grown plants for a week. Plant seedlings out in summer when they have reached a height of about 8cm (3in). Water thoroughly and gently lift the seedlings from the soil using a hand fork. Space plants 60cm (2ft) apart in every direction, planting into shallow holes and firming soil around the roots.

Growing broccoli under fine mesh will minimize insect damage.

PROTECT Place barriers and traps around young plants to provide protection from slugs and snails (*see p.38*). Cover the seedlings with fine insect mesh to keep off flea beetles and place collars (see *p.135*) around the base of young plants to prevent cabbage root flies laying their eggs. Use netting to keep off egg-laying butterflies in summer and pigeons throughout growth. Stake plants at risk of blowing over.

WATER AND FEED Water regularly while the transplanted plants establish and also during dry summer weather. Boost growth with a balanced liquid feed in late summer. Keep the surrounding soil free of weeds. Add a mulch of compost around plants in autumn, both to improve the soil and suppress weeds.

Water sprouting broccoli growing in free-draining raised beds regularly.

HARVEST

Broccoli forms its edible flowering shoots from early winter to mid-spring. Snap or cut stems when 10–15cm (4–6in) long, well before the buds open, starting with the central shoot. Harvest the shoots regularly to promote further growth; a single plant can yield shoots for 6–8 weeks. Harvest tenderstem varieties in the same way from midsummer to mid-autumn.

Cut the central spear first to promote side shoots. To maximize yield, pick evenly from several plants rather than stripping one at a time.

WHY NOT TRY?

If you don't have the time or space for sprouting broccoli, then try broccoli raab, which can be ready to cut just 7–8 weeks after sowing. The plants are also much smaller, reaching just 30cm (12in) tall. Sow thinly, 1cm (½in) deep, directly into the soil or large containers, every three weeks from early spring to midsummer. Thin to 15cm (6in) apart and cut the flowering stems above the lowest leaves when they reach about 20cm (8in) tall. Plants will resprout several times.

The leafy flower stems of broccoli raab have a mustard-tinged flavour.

VARIETIES

Grow several varieties for a longer cropping season. White varieties are less hardy than those with purple buds.

'BURBANK' An elegant white variety that is ready to pick from mid-winter.

'CLARET' A vigorous, late, purple variety with plentiful thick stems to harvest in spring.

'INSPIRATION' This tenderstem broccoli yields quick summer crops.

'RED ADMIRAL' Try this extra-early variety for tender dark purple spears from early winter.

'RED ARROW' Reliable and prolific, this variety sends up purple shoots over a long season from mid-winter.

SPRING CABBAGE

With its distinctive pointed heads of mildly flavoured, nutritious leaves, spring cabbage provides a welcome harvest in mid-spring. It is ideal for planting in beds in autumn, after the summer vegetables are harvested, and is hardy enough to overwinter for spring greens or a headed crop.

DIFFICULTY Easy
WHEN TO SOW Late July to Aug
IDEAL SOIL TYPE Fertile, well-drained, and firm
SITE REQUIREMENTS Full sun
GERMINATION TIME 4–7 days
GROW FROM Seeds or plants
YIELD 6–10 heads per 2m (6ft) row

CALENDAR

	WINTER			SPRING			SUMMER			AUTUMN		
SOW								Spring greens / Headed cabbage	Spring greens / Headed cabbage			
HARVEST				Spring greens	Headed cabbage	Headed cabbage	Headed cabbage					

Spring greens
Headed cabbage

Time between sowing and harvesting
8 months

SOWING

Spring cabbages are usually sown in a seedbed or in modules from late July (in cold areas) to August (in milder regions). They are then planted out in their final positions in autumn. Young plants are also readily available from garden centres or by mail order.

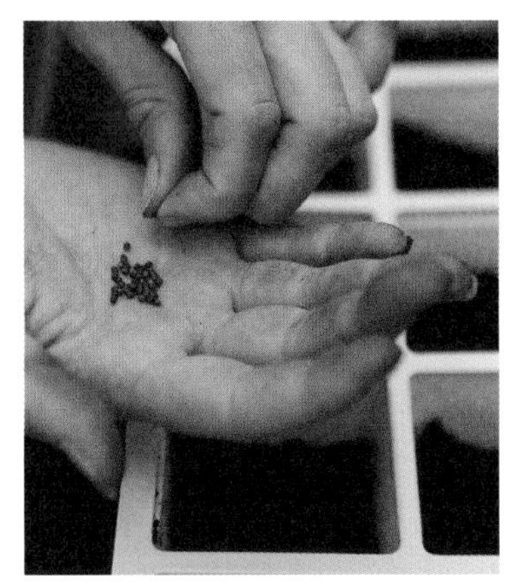

Cabbages sown in modules will be ready to plant out in about four weeks.

SOWING INTO THE SOIL To raise plants for transplanting, find space for a row 60–90cm (2–3ft) long. Rake the soil in this seedbed to a fine, crumbly texture and, using a trowel, create a trench 1cm (½in) deep. Sow the seeds thinly along the row, cover with soil, water thoroughly, and label. Where there is room in the garden, spring cabbages can also be sown in their final position. In this event, sow them thinly along rows 30cm (12in) apart.

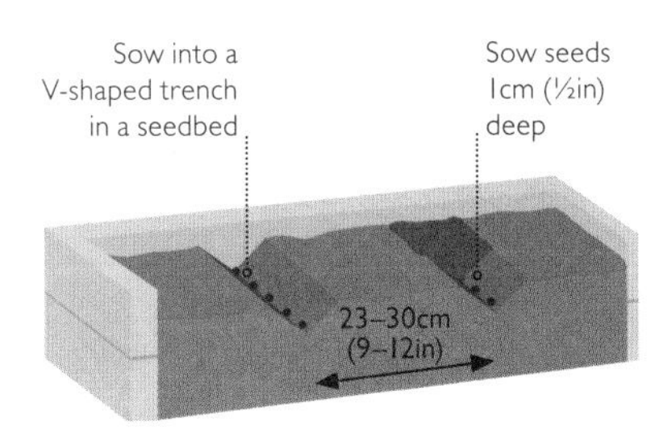

SOWING IN MODULES Sow single seeds 1cm (½in) deep into the centre of modules filled with multi-purpose compost. Cover them with compost, water, label, and let them germinate on a cool, bright windowsill or in a sheltered position outdoors.

GROW

THIN Thin seedlings sown in a seedbed promptly to 5cm (2in) apart. Those sown in their final position can gradually be thinned to 10cm (4in) apart. Weed around the rows regularly.

TRANSPLANT Move young plants from seedbeds or modules into the ground from early to mid-autumn, when they have four leaves. Choose a sunny site with fertile, slightly alkaline soil (*see p.20*). Harden off plants that you have raised indoors. Water well before pushing plants out of their modules or carefully lifting those grown in the soil with a hand fork. Plant out in rows 30cm (12in) apart, at 30cm (12in)

Plant out or transplant on a cool, cloudy day to prevent the plants wilting.

intervals for headed cabbages, or 10cm (6in) apart if you want spring greens first and headed cabbages later.

PROTECT Rotate cabbage (see *p.23*) to minimize damage from diseases and pests. Grow young plants under fleece or mesh to protect against cabbage root fly and flea beetles; root flies will also be deterred by the use of "cabbage collars" – discs that shield the soil at the base of the plant. Create barriers and traps around plants to control slugs and snails. Netting stretched over supports will keep hungry pigeons away.

Cabbage collars can be bought from a garden centre or cut from cardboard.

HARVEST

Lift young plants, or cut their stems with a knife, for greens in early spring, leaving one plant in three to heart-up.

The first headed cabbages should be ready to harvest in mid-spring. Cut the heads with a sharp knife; they are delicious boiled, braised, or cut into thick wedges and roasted.

Hold the large leaves aside and use a sharp knife to cut the thick, woody stem beneath headed spring cabbages.

WHY NOT TRY?

Double cropping is a technique that allows a single plant to produce two harvests and works well for spring cabbages. Harvest the cabbage by cutting its stalk with a knife, leaving the two lowest leaves attached. Use a knife to cut a shallow cross into the top of the stem, which will then start to develop new shoots. Water these stumps regularly and feed with a high-nitrogen fertilizer once the new shoots begin to grow. There could be as many as five new small heads of spring greens to pick within about eight weeks.

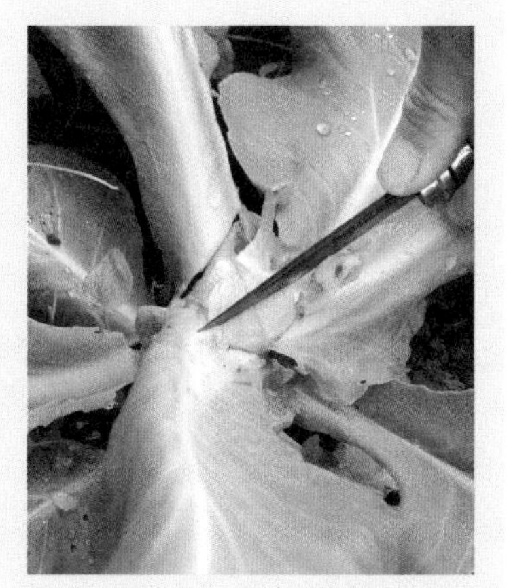

Cut a cross into the cut cabbage stump to promote a new crop.

VARIETIES

Size is the main differentiating factor between spring cabbage varieties, so select one to suit your plot.

'ADVANTAGE' This medium-sized variety yields early spring greens and tight heads which stand well once mature.

'APRIL' Perfect for spring greens, this compact variety is quick to heart-up.

'DURHAM EARLY' Hardy and reliable, this variety produces an early crop of spring greens and small, pointed cabbages.

'HISPI' The medium to large pointed heads are of this variety are packed with full-flavoured leaves.

'SPRING HERO' A rounded, ball-head variety – rare among spring cabbages.

KALE

These large, handsome plants are extremely hardy and produce colourful, textured, and highly nutritious leaves. Traditionally, kale is harvested through winter into spring, but plants grown close together in open soil or containers yield a delicious summer and autumn crop of tender leaves.

DIFFICULTY Easy
WHEN TO SOW Apr to June
IDEAL SOIL TYPE Well drained and fertile
SITE REQUIREMENTS Full sun or light shade
GERMINATION TIME 7–14 days
GROW FROM Seeds or plants
YIELD 3kg (6½lb) per 2m (6ft) row

CALENDAR

	WINTER	SPRING	SUMMER	AUTUMN
SOW				
HARVEST				

Early crop
Main crop

Time between sowing and harvesting
3–6 months

SOWING

Kale plants are usually sown into a seedbed in open soil, or in modules or small pots to transplant to their final positions in summer. Five productive plants are often sufficient in a small garden. Early crops to harvest in summer can be sown directly into the soil or in large containers.

SOW IN THE SOIL Raise young plants for transplanting in a short row 60–90cm (2–3ft) long. From mid-spring to early summer, make a drill 1cm (½in) deep with a cane or trowel and sow seeds thinly along it. Cover with soil, water, and label the row. Thin seedlings to 8cm (3in) when they are big enough to handle.

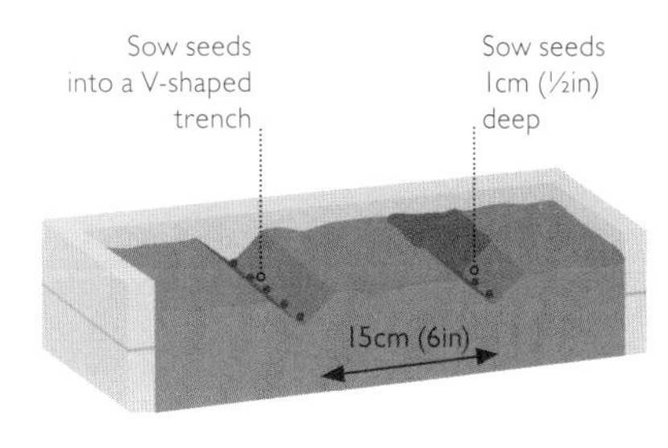

You can also sow early crops directly into their final positions from mid- to late spring, in rows 15cm (6in) apart.

SOW IN POTS From mid-spring, fill small pots or modules with compost. Sow one seed 1cm (½in) deep in the centre of each. Cover with compost, water, label, and place in a sheltered spot outdoors or on a bright windowsill.

Plants sown in small pots can be transplanted about eight weeks later.

Plant out young kale into open soil when plants have five or six leaves.

GROW

TRANSPLANT Plants originally sown in small pots or modules may need potting on into larger containers before they are ready to transplant into their final positions in early to midsummer. Plant them into well-drained, fertile, slightly alkaline soil (*see p.20*). Space plants 45cm (18in) apart each way for main overwintering crops or 15cm (6in) apart each way for summer leaves. Dig a hole deeper than the rootball, position each plant with its lowest leaves just above the soil surface, and firm the soil around the roots. Thin direct-sown seedlings of early kale to 15cm (6in) apart.

WATER AND FEED Water the seedlings thoroughly after transplanting and keep the soil consistently moist during summer. Weed around plants regularly. Mulch in autumn to retain soil moisture and suppress weeds. Feed plants with a high-nitrogen fertilizer in early spring to boost growth.

PROTECT Cover plants with fine netting or fleece to keep off egg-laying butterflies and pigeons.

Kale is fairly problem-free, but netting is a sensible precaution.

Kale leaves are ready to harvest when they are about the width of an adult's hand.

HARVEST

Start harvesting early crops of summer kale three months after sowing, and main winter crops in late autumn. Cut or snap the leaves from the base of the stem upwards and pick evenly across all of your plants. Kale will continue to grow slowly from the top of the leaf cluster during mild winter weather and some varieties put out tender side-shoots in spring. Stop picking when flower stems appear and the leaves toughen. Chop small leaves raw into summer salads; larger leaves are lovely steamed, stir-fried, or added to soups and pasta dishes.

WHY NOT TRY?

A cross between kale and Brussels sprouts, kalettes produce small, sprout-sized clusters of purple-tinged leaves from a sturdy, upright stem. They need a fertile, well-drained soil. Sow in early to mid-spring 1cm (½in) deep in a seedbed or modules, and transplant to 50cm (20in) apart in late spring or early summer. Keep soil moist during summer and pull the soil up around stems to increase stability. Pick from the base of the stem upwards, from late autumn until early spring.

Kalettes are tall plants that need shelter from strong winds

VARIETIES

Kale varieties come in an array of leaf colours and textures. Try dwarf varieties for small gardens.

'DWARF GREEN CURLED' A compact curly kale with dark green, frilled leaves.

'EMERALD ICE' Broad, white-veined leaves are strikingly curled at the edges.

'NERO DI TOSCANA' Also known as cavolo nero or black kale, this delicious Italian variety has long, blistered, dark leaves.

'REDBOR' Incredibly ornamental, rich purple, curled leaves look wonderful in the winter garden.

'STARBOR' This dwarf, green, curly kale is ideal for small plots and containers.

ORIENTAL GREENS

The cabbage family includes a variety of hardy, leafy Asian crops such as mizuna, mibuna, mustard greens, pak choi, tatsoi, and komatsuna. Leaves range in flavour from mild and cabbage-like to fiery mustard. They do best sown in summer to harvest through autumn and winter.

DIFFICULTY Easy
WHEN TO SOW June to Aug
IDEAL SOIL TYPE Well drained, moist, and fertile
SITE REQUIREMENTS Full sun or light shade
GERMINATION TIME 3–10 days
GROW FROM Seeds
YIELD 1.5kg (3lb) per 2m (6ft) row

CALENDAR

	WINTER			SPRING			SUMMER			AUTUMN		
SOW												
HARVEST												

Time between sowing and harvesting
8–10 weeks

SOWING

Oriental greens are prone to bolting from spring sowings, which makes them tricky to grow to maturity. Instead, try making two or three sowings from early to late summer for a succession of leaves from late summer into winter. These plants often do well when sown directly into the soil, but raising them in modules is worthwhile where slugs are prevalent or you are growing to replace a crop that will be removed in early autumn, such as runner beans.

Oriental leaves are generally hardy and can be sown directly into the soil.

Pak choi seedlings need thinning while they are small to achieve a good crop.

SOWING INTO THE SOIL Improve the soil with plenty of well-rotted compost before planting to help retain moisture. From early to late summer, stretch a string line across the bed and create a trench 1cm (½in) deep with a cane or trowel. Sow seed thinly along the row, cover with soil, label the row, and water thoroughly. Space rows 23–30cm (9–12in) apart.

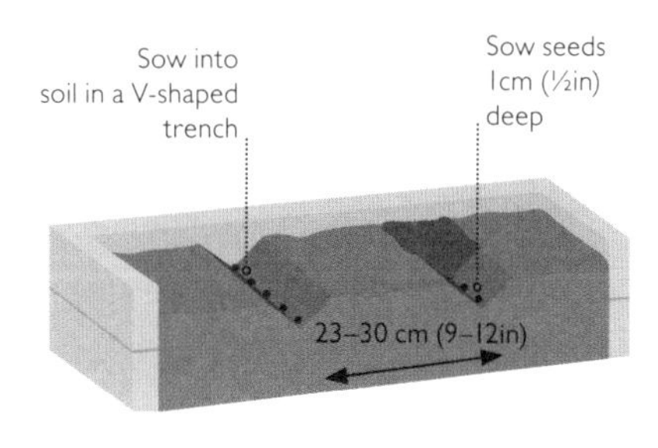

SOWING INTO MODULES Fill a module tray or small pots with multi-purpose compost and sow a single seed, 1cm (½) deep, into the centre of each container. Cover with compost, water well, label, and place in a sheltered spot outdoors or a cool windowsill.

GROW

THIN OR TRANSPLANT Thin these fast-growing seedlings promptly to 15cm (6in) apart if you wish to harvest smaller leaves, or 30cm (12in) apart if you want fully mature plants that produce larger leaves over a long period. Add the thinnings to salads. Transplant young plants grown in modules into their final positions in the soil or large containers 3–4 weeks after sowing, at the same spacings in rows 23–30cm (9–12in) apart.

WATER All oriental greens are prone to bolting in hot, dry conditions, so water young plants regularly during summer to keep the soil moist. Perhaps the best solution to bolting is simply to eat the leaves young. Keep the soil free of weeds by hoeing or weeding by hand every two weeks.

Oriental greens are fairly shallow-rooted and need regular watering.

PROTECT Slugs and snails can develop a taste for these seedlings, so set barriers and traps around plants. Sowings made later in summer should escape flea beetle attacks (*see p.140*), but cover early summer sowings with fleece or insect mesh to keep them out. Although these plants are hardy, they can be spoiled by cold or wet winter weather, and should be covered with a cloche or fleece tunnel from early winter.

Garden fleece helps protect early pak choi from flea beetle attack.

HARVEST

Plants will be ready to harvest from about eight weeks after sowing, when about 15cm (6in) tall. For a continuous crop and well-developed plants to overwinter, pick the largest leaves from the outside of plants as required.

Alternatively, cut the entire head of leaves about 2.5cm (1in) above ground level for pak choi, tatsoi, and komatsuna, and 5cm (2in) above the soil for mizuna, mibuna, and mustard greens, and allow plants to resprout. There will be little growth over winter, but a flush of new leaves is often produced in early spring.

Mustard greens and other oriental leaves are often grown alongside salad plants.

VARIETIES

Oriental leaves vary widely in appearance, flavour, and texture.

KOMATSUNA Spoon-shaped leaves have a mild mustard flavour; plants are less prone to bolting than many oriental greens.

MIBUNA Long, narrow, green leaves form attractive, fast-growing clumps.

MIZUNA 'KYOTO' Clumps of white-stemmed, feathery, green leaves have a gentle tang of mustard.

MIZUNA 'RED KNIGHT' Hardy, rounded, burgundy leaves have a peppery flavour.

MUSTARD 'RED FRILL' Dark red, finely divided leaves pack a punch of mustard.

PAK CHOI 'JOI CHOI' A hardy variety, with thick white stems and glossy, dark green leaves.

PAK CHOI 'RUBI' Superb rosettes of rich claret leaves contrast with pale green stems.

TATSOI 'ROZETTO' This beautiful, rosette-forming type of pak choi is very hardy.

TROUBLESHOOTING

The lush foliage of leafy veg needs protection, because it is a magnet for slugs, snails, and pigeons, which can strip mature plants fast in winter. Members of the cabbage family are targeted by specific pests and diseases; fortunately crop rotation, preventative measures, and protective barriers all provide effective ways to keep them healthy.

CABBAGE ROOT FLY

PROBLEM Plants in the cabbage family grow poorly, wilt, and die.
CAUSE Adult cabbage root flies lay eggs on the soil around plants, which hatch into root-eating maggots.
REMEDY Protect crops under fleece or fine insect mesh barriers, or by placing collars of cardboard or other sturdy material around the stem of each plant.

CABBAGE WHITEFLY

PROBLEM Small white insects on the undersides of leaves on cabbages, kale, and sprouting broccoli.
CAUSE The sap-sucking insect, cabbage whitefly.
REMEDY This pest is usually tolerable, but appropriate pesticides can be effective when applied several times. Check the manufacturer's guidelines.

CATERPILLARS

PROBLEM Areas of foliage are eaten away on plants of the cabbage family during summer.
CAUSE Caterpillars of butterflies and moths feeding on leaves.
REMEDY Manually pick off any tiny yellow or white eggs or caterpillars. Stretch fine netting over supports above plants, so that insects can't lay eggs through it.

FLEA BEETLE

PROBLEM Small, round holes in leaves of crops in the cabbage family, which may stunt the growth of seedlings.
CAUSE Tiny flea beetles, which can often be seen jumping from disturbed foliage.
REMEDY Sow when the soil is warm and water well to encourage rapid growth, or start seedlings in modules. Cover rows with fine insect mesh after sowing.

MEALY CABBAGE APHIDS

PROBLEM Leaves of cabbages, kale, and sprouting broccoli develop pale yellow patches and can become distorted.
CAUSE Grey-white, sap-sucking aphids on the undersides of leaves.
REMEDY Small numbers can be tolerated; squash any you find between thumb and forefinger. Encourage predatory insects by planting flowering plants alongside.

SLUGS AND SNAILS

PROBLEM Large, ragged holes in leaves; the foliage and stems of seedlings are badly damaged overnight.
CAUSE Slugs and snails emerging from damp and cool hiding places to feed.
REMEDY Remove suitable hiding places near the veg patch and collect the culprits on torchlit patrols at night. Place beer traps and barriers around crops.

WOOD PIGEONS

PROBLEM Leaves have angular tears or are stripped down to the stems.
CAUSE Wood pigeons, which are especially hungry in winter and early spring.
REMEDY Make a sturdy frame, stretch netting over it, and secure it to the ground. Pigeons may land on loose netting and peck through it.

BOLTING

PROBLEM Crops send up flowering shoots prematurely, halting leaf production.
CAUSE Sowing in the wrong season or when conditions are too cold. Dry soil also triggers bolting.
REMEDY Follow sowing advice and wait for the soil to warm. Thin promptly to prevent competition for moisture and water regularly during warm weather.

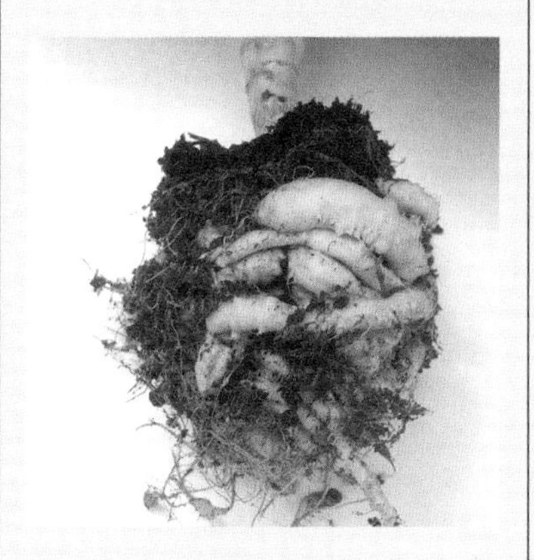

CLUB ROOT

PROBLEM Plants in the cabbage family grow poorly. Leaves may turn purple and wilt in hot weather. Roots are swollen and distorted.
CAUSE The fungal infection club root.
REMEDY Improve drainage and lime acid soils to raise the pH. Be careful not to buy or accept infected plants. Practice crop rotation.

YELLOWING LEAVES

PROBLEM The lower leaves of kales, cabbages, and sprouting broccoli yellow and fall in late winter and early spring.
CAUSE A natural drop of old leaves.
REMEDY There is no cause for concern; remove and compost yellowing foliage. Where new growth shows signs of yellowing, feed with a high-nitrogen fertilizer and check for pests.

INDEX

Bold text indicates a main entry for the subject.

A

allotments 49
amaranth 61, 129
aphids 39, 66, 80, 110, 111, 141
artichokes
  globe **108–09**
  Jerusalem 109
aubergines 48, **104**
  blossom end rot 111
  planting out 31
  season extension 45

B

basil 59, **62–63**
beans *see* peas and beans
beetroot **116**
  beet leaf miner 124
  extending season 44
  planting out 31
  storage 40
biological control 38
birds 29, 92
  beneficial 39
  pigeons 19, 38, 71, 81, 141
bolting 67, 125, 141
borlotti beans **78–79**
broad beans **76–77**
  chocolate spot 81
  companion planting 39
  planting out 31
broccoli see sprouting broccoli

C

cabbage family
  crop rotation 23
  double cropping 135
  kale see kale
  Oriental greens 31, **138–39**
  ornamental planting 49
  pests and diseases 66, 140, 141
  planting out 31, 134, 136
  second season 43
  spring cabbage **134–35**
callaloo 129
carrots **114–15**
  carrot fly 39, 66, 115, 124
  companion planting 39
  extending season 44
  storage 40
caterpillars 81, 92, 140
chard **130–31**
chervil **62–63**
chicories see radicchio and curly endive
children **26–27**
chillies 48, **105**
  planting out 31
  season extension 45
  storage 40
  watering 37
  see also peppers
clay soil, dealing with 21
cloches 30, 45
community gardens 49
companion planting 39
composting 9, **46–47**
  exceptions 47
  see also feeding plants; organic gardening
containers 15, **18–19**, 24, 26, 29
  compost and feeding 19, 35
  pests and diseases 19, 38
  potatoes 121, 122
  watering 37
coriander 61, **62–63**
courgettes **100–101**
  diseases 111
  liquid feeds 35
  planting out 31
cress and microgreens **59**
crop rotation 23
cucamelons 103
cucumbers **102–03**, 111
curly endive see radicchio and curly endive
cut-and-come-again leaves see lettuce; peppery leaves

D

damping off 33, 67, 93
dill **62–63**

E

edamame beans 79

F

family gardening **26–27**
feeding plants **34–35**
  post-harvest green manure 41
  see also composting; organic gardening
fennel (leafy herb) **64–65**
fertilizers 34, 35
  see also composting
flea beetle 57, 66, 124, 140
fleece tunnels 38, 45
follow-on crops 42
French beans 31, **72–73**

G

garlic **88**
  pests 92, 93
  storage 40, 88
globe artichokes **108–09**
gluts, dealing with 43
greenhouses 33

H

hardening off 30
harvesting **40–41**
herbs see salads and herbs

I

intercropping 23, 42

J

Jerusalem artichokes 109

K

kale 49, **136–37**
  as microgreen 61
  planting out 31, 136
  successional sowing 44, 136
  yellowing leaves 141
kohlrabi 49, **123**
  flea beetle 124
komatsuna 57, **138–39**

L

leaf beet **130–31**
leaf celery **59**
leafy herbs **62–65**
leafy veg *see* cabbage family; chard; spinach
leeks **90–91**
  pests and diseases 92, 93
  planting out 31, 91
lettuce **52–53**
  pests and diseases 52, 66, 67
  planting out 31

M

mangetouts **70–71**
mibuna **138–39**
mice, and peas and beans 80
microgreens **59**
mint **64–65**, 67
mizuna **138–39**
mustard greens 57, 61, 118, **138–39**

N

nematodes 38
no-dig methods 21

O

onion family 39, 82–93
  bolting 93
  damping off 93
  garlic see garlic
  leeks see leeks
  onions **84–85**
  pests and diseases 92, 93
  planting out 31, 91
  shallots **86**
  spring onions **87**
  storage 40, 88
  tree onions 85
organic gardening 9, 20, 21, 35
  see also composting; feeding plants
Oriental greens 31, **138–39**
ornamental planting 49
overwatering 37

P

pak choi 57, **138–39**
pallet garden 25
parsley **62–63**, 66
parsnips **119**
  pests and diseases 39, 115, 124, 125

peas and beans 48, 68–81
bean and pea shoots **60**, 71
borlotti beans **78–79**
broad beans *see* broad beans
drying beans **78–79**
edamame beans 79
extending season 44
French beans 31, **72–73**
mangetouts **70–71**
peas **70–71**
pests and diseases 75, 78, 80, 81
planting out 31
runner beans **74–75**
setting problems 81
stakes and supports 71, 72–73, 74–75, 77, 79
storage 40
successional sowing 44
sugar snap peas **71**
peppers **105**
blossom end rot 111
chillies *see* chillies
planting out 31
season extension 45
peppery leaves **56–57**, 138
flea beetle protection 57
as microgreens 61
see also radicchio and curly endive
pests and diseases
cabbage family 66, 140–41
containers 19, 38
onion family 92–93
peas and beans 75, 78, 80–81
root veg 39, 66, 115, 124–25
salads and herbs 55, 57, 66–67
summer veg 110–11
pigeons 19, 38, 71, 81, 141
see also birds
planning and planting 6–7, **14–15**, 22
raised beds 15, **16–17**
second season 42, 43
small spaces 24
planting out 31
potatoes **120–22**
in containers 121, 122
diseases 125
extending season 44, 120, 122
powdery mildew 67, 81, 125
productivity boosting **22–23**
propagators 32, 33
protection
barriers for 38
cloches 30, 45
raising plants under cover **32–33**
seeds 29
small spaces 25

R

rabbits 38
radicchio and curly endive **54–55**
blanching 55
see also peppery leaves
radishes **117**
flea beetle 124
successional sowing 44, 117
rainwater 37
raised beds 15, **16–17**, 48
raising plants under cover **32–33**
record keeping 41
rocket *see* peppery leaves
root vegetables *see* beetroot; carrots; kohlrabi; potatoes; turnips
rosemary 65
runner beans **74–75**
planting out 31
storage 40

S

sage 65
salads and herbs 49, 50–67
basil 59, **62–63**
bolting 67
chervil **62–63**
chives **89**
coriander 61, **62–63**
damping off 67
dill **62–63**
leaf celery **59**
leafy herbs **62–65**
lettuce *see* lettuce
mint **64–65**, 67
parsley **62–63**, 66
pea and bean shoots **60**, 71
peppery leaves **56–57**
pests and diseases 55, 57, 66–67
radicchio and curly endive **54–55**
season extension 45
sorrel **58**
storage 40
successional sowing 44
season extension **44–45**
seasons and timing 11, **12–13**
second season **42–43**
seed, starting from **28–29**
shallots **86**
pests and diseases 92, 93
storage 40
slugs and snails 31, **38**
leafy vegetables 141
peas and beans 75, 78, 81
root vegetables 115, 124
salads and herbs 55, 67
small spaces **24–25**
vertical space use 24, 101
soil preparation **20–21**
sorrel **58**
sowing seeds **28–29**
space, freeing up 10, **48–49**
spinach **128–29**
New Zealand 131
storage 40
successional sowing 44
spring greens, broad bean tops 77
spring onions **87**
sprouting broccoli 23, **132–33**
pests and diseases 140, 141
planting out 31
squashes 48, **100–101**
cucumber mosaic virus 111
planting out 31
stakes and supports 71, 72–75, 77, 79
storage 40
successional sowing 43, 44
sugar snap peas **71**
summer veg *see* artichokes; aubergines; courgettes; cucumbers; squashes; sweetcorn; tomatoes
sweetcorn 49, **106–07**
planting out 31
popping corn 107
Swiss chard 57, 61

T

tarragon **64–65**
tatsoi **138–39**
thyme 65
timing
and seasons 11, **12–13**
watering 36, 37
tomatoes
bush tomatoes **96–97**
cordon tomatoes **98–99**
diseases 111
as edible bedding 48
extending season 44
fruit splitting 111
liquid feeds 35
planting out 31
storage 40
watering 37
tree onions 85
troubleshooting *see* pests and diseases
turnips **118**
flea beetle 124
storage 40
two-in-one crops 23

U

underplanting 23

V

vertical space use 24, 101

W

watering **36–37**
timing 36, 37
water butt 37
weeding 21
wildlife, attracting beneficial 39

Author Jo Whittingham

PUBLISHER ACKNOWLEDGMENTS

DK would like to thank Oreolu Grillo and Sophie State for early spread development for the series, and Margaret McCormack for indexing.

PICTURE CREDITS

The publisher would like to thank the following for their kind permission to reproduce their photographs:

Alamy Stock Photo: Mark Bolton Photography 6c; Nigel Noyes 13bc; Tim Gainey 14c; Tim Gainey 19cl; 24bl; 25tc; Peter Jordan_NE 30bl; Virginija Vaidakavičienė 35t; BIOSPHOTO 38bc; Anne Gilbert 41bl; Art Directors & TRIP 48bl; Christopher Nicholson 49br; Dorling Kindersley Ltd 52c; Dave Bevan 66bl; Thomas Smith 72bc; Alison Thompson 73tl; Clare Gainey 75bc; GKSFlorapics 76c; John Swithinbank 77cl; Clare Gainey 79tc; Nigel Cattlin 80tr; Nigel Cattlin 81tr; GKSFlorapics 82c; Alison Thompson 91bc; Graham Corney 91br; tbkmedia.de 92bl; Nigel Cattlin 92br; Tomasz Klejdysz 93tl; Matthew Taylor 93tc; Nigel Cattlin 93bc; Tim Gainey 97br; keith morris 109tr; Nigel Cattlin 110bc; Nigel Cattlin 111bl; Nigel Cattlin 111bc; Dorling Kindersley ltd 114br; Alicia Neumiler 115cr; Manfred Ruckszio 124br; Avalon/Photoshot License 125tc; Nigel Cattlin 125tr; Nigel Cattlin 125bc; Denis Crawford 125br; Kathy deWitt 137bc.

Dorling Kindersley: Peter Anderson 107cr;

GAP Photos: 22bl; 23tr; Elke Borkowski 24br; BBC Magazines LTD 32bc; 33tl; 33tc; 33tr; Jonathan Buckley 56br; Friedrich Strauss 57tr; 60c; 60cr; John Swithinbank 106bl; Jo Whitworth 107br; Friedrich Strauss 108c; FhF Greenmedia 135bc.

Getty Images: Dougal Waters 2c; Fabian Krause/EyeEm 3l; cjp 15cl; Helin Loik-Tomson 32tl; Klaus Vedfelt 36tl; stevanovicigor 41tr.

Huw Richards: 12bl; 17tl; 17tr; 17tcl; 17tcr; 17bcl; 17bcr; 17b; 20cr; 26tr; 26br; 46cl; 46bl; 46br; 84c; 85tr; 129cl; 129tr.

Illustrations by Cobalt id

Produced for DK by
COBALT ID

Managing Editor Marek Walisiewicz
Managing Art Editor Paul Reid
Art Editor Roger Walton

DK LONDON
Project Editor Amy Slack
Managing Editor Ruth O'Rourke
Managing Art Editor Christine Keilty
Production Editor David Almond
Production Controller Stephanie McConnell
Jacket Designer Nicola Powling
Jacket Co-ordinator Lucy Philpott
Art Director Maxine Pedliham
Publishers Mary-Clare Jerram, Katie Cowan

First published in Great Britain in 2021 by
Dorling Kindersley Limited
DK, One Embassy Gardens, 8 Viaduct Gardens,
London, SW11 7BW

10 9 8 7 6 5 4 3 2 1
001- 320997-Mar/2021

A CIP catalogue record for this book is available from the British Library.
ISBN: 978-0-2414-5859-4

Printed and bound in China

For the curious
www.dk.com